BEGINNING ALGEBRA
for College Students

Nancy Myers
BUNKER HILL COMMUNITY COLLEGE

MERRILL PUBLISHING COMPANY
A BELL & HOWELL INFORMATION COMPANY
COLUMBUS   TORONTO   LONDON   MELBOURNE

Cover Illustration: George Gruber
Cover Designer: Cathy Watterson

Published by Merrill Publishing Company
A Bell & Howell Information Company
Columbus, Ohio 43216

This book was set in Times Roman.

Administrative Editor: John Yarley
Production Coordinator: Rex Davidson
Art Coordinator: Pete Robison
Text Designer: Connie Young

Library of Congress Catalog Card Number: 88–061052
International Standard Book Number: 0–675–20924–2
Printed in the United States of America
1 2 3 4 5 6 7 8 9 — 93 92 91 90 89

Preface

Beginning Algebra for College Students is written for students who have never studied algebra, or who need to review the basic concepts of algebra. It is assumed that students have some mastery of arithmetic, including the arithmetic of whole numbers, fractions, and decimals. Absolutely no prior knowledge of variables, equations, or signed numbers is assumed.

Beginning Algebra for College Students was originally written for use in self-paced courses; however, the text is also appropriate for lecture courses. The essentials of a lecture are built into each section, review topics are covered as needed, and longer proofs are left to the ends of units.

Other features of the text include:

- *A unit form of organization*: This allows the student to master a single concept in each section and a limited number of related topics in each unit. A unit is usually less than half the length of an average chapter, covering two to five closely related objectives. Each section within a unit covers just one topic described by one objective.
- *Cross-referencing*: The entire text is cross-referenced by objective numbers. At the beginning of each unit there are at least two but no more than five objectives. Each section within the unit corresponds by number to an objective. Each item in the self-test at the end of the unit is keyed to an objective number by a reference in the answer section. Thus, an incorrect answer on a self-test leads the student by number to the appropriate objective, and then by number to the appropriate section for further study.
- *Worked examples*: The worked examples are explained with full sentences preceding and between steps, establishing a dialog with the student. As the examples progress in a section, the amount of explanation, especially for steps done in previous examples, gradually decreases, encouraging students to fill in their own explanations.
- *Exercise sets*: Every section has an exercise set. Since the sections are generally shorter than standard chapters and sections, exercise sets occur more frequently, but they are purposefully made shorter than standard sets. The student works on a new concept right away, on only one concept at a time, and with a list of problems that is not so long as to be discouraging. Additionally, the first several exercises in each set closely follow the worked examples, the first two or four matching the first example, the next couple matching the second example, and so on. Odd and even exercises are paired, and answers to all exercises are given at the end of the text.
- *Self-tests*: Each unit ends with a self-test consisting of five items. The items may be in random order, but the answers are keyed to objectives and sections as previously described.
- *Cumulative Reviews*: After each four or five units there is a cumulative review unit. The self-tests in cumulative review units contain ten items with answers at the end of the text. The answers are keyed to a unit, but not to a specific section.

A test bank to accompany *Beginning Algebra for College Students* is available to instructors. It contains a pretest, four forms of unit tests with five items each, four forms of cumulative review tests with ten items each, and two forms of final exams.

The material in this text falls into three parts. The first part, consisting of Units 1 through 8, introduces the basic concepts of algebra, including variables, expressions, equations, and signed numbers. A standard course will cover all, or almost all, of the material in this part.

A second part, consisting of Units 9 through 17, extends the material to other algebraic forms, including formulas, graphs, systems of equations, and inequalities—all for linear forms. This part also introduces higher orders through exponents and polynomials. The material in this part is prerequisite to a second-level course in algebra and has been motivated by the first part. Most courses will cover the bulk of this material.

The third part, Units 18 through 21, uses the forms studied in the second part to provide a preview of expressions and equations at the next level, including rational expressions, equations involving rational expressions, and quadratic equations. Instructors can use their discretion in selecting topics from these units to complete a one-semester course.

I wish to thank my colleagues in the mathematics department of Bunker Hill Community College who have used my texts in their classrooms and who have provided support and encouragement, as well as valuable feedback. These colleagues include Robert Carlson, Charles Chisholm, Herbert Gross, Adele Hamblett, Robert Leonard, Shirley MacKenzie, Joanne Manville, Joan Namvar, Dorothy Ryan, Yvette Straughter, Maurice Temple, and Judith Tully, numerous part-time faculty members, and former colleagues Virginia Campiola and Joan McGowan. Maria Trenga, a former student, typed the manuscript. And countless students helped with their comments and questions.

I also wish to thank Sheila B. Cox, Clinch Valley College and Carl T. Carlson, Moorehead State University who reviewed the manuscript. Their critical comments and thoughtful input helped guide me in writing this text.

Finally, I thank John Yarley, Pete Robison, and Rex Davidson of Merrill Publishing Company for their assistance in producing this text.

Nancy Myers

Contents

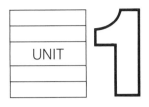

UNIT 1

Algebraic Expressions

INTRODUCTION

Algebra and arithmetic are closely related. Algebra might be said to be a generalization of arithmetic. Arithmetic uses numbers and operation symbols (plus, minus, etc.). In algebra, letters may be used in place of numbers. In this unit you will learn how letters may replace numbers. You will also review some basic properties of arithmetic that are basic properties of algebra as well.

OBJECTIVES

When you have finished this unit you should be able to:

1. Identify the variables and constants, the number of terms, and the numerical coefficient of a variable in a term of an algebraic expression.
2. Evaluate an algebraic expression, given numerical values for the variables.
3. Identify examples of the associative and the commutative properties for addition and for multiplication.
4. Use the distributive property for multiplication over addition to remove parentheses in an algebraic expression.

1.1 Constants and Variables

The word *algebra* derives from the word *al-jabr* in the title of a book written by an Arabian mathematician, Al-Khowârizmî, in the ninth century. The full title of the book is *Hisâb al-jabr w'al-muqâbalah*. It was translated into Latin and introduced into Europe in the twelfth century. The word *al-jabr* means ''restoring,'' as in restoring the numerical value of a letter. We will see the elements of this process in Unit 2.

The basic component of algebra is the **algebraic expression.** An algebraic expression represents a number. In arithmetic, you have used numbers and operation symbols to represent numbers. For example,

$$7 + 9 \text{ represents } 16$$

since the sum of 7 and 9 is 16. Also,

$$5 \times 11 \text{ represents } 55$$

since the product of 5 and 11 is 55. Similarly,

$$25 - 6 \text{ represents } 19$$

and

$$56 \div 8 \text{ represents } 7.$$

These expressions are arithmetic statements, but they also are algebraic expressions since each represents a number.

Algebraic expressions that are not arithmetic statements include letters as well as numbers and operation symbols. For example,

$$7 + x$$

is an algebraic expression. The addition

$$x + y$$

is also an algebraic expression. Similarly,

$$25 - x$$

and

$$x - y + z$$

are algebraic expressions.

Expressions such as

$$7 + x, \quad x + y, \quad 25 - x, \quad x - y + z$$

may not appear to represent numbers because of the presence of the letters x, y, and z. However, the letters themselves represent numbers. In these expressions, x, y, and z may represent any number. Since they may vary among numerical values, x, y, and z are called **variables.** The numbers 7 and 25 can represent only 7 and 25. They constantly represent the same number, so they are called **constants.**

EXAMPLE 1.1   Identify the constants and the variables in each of these algebraic expressions:

a. $y + 13$ b. $s + t$

Solutions

a. Since numbers are always constants, the number 13 is a constant. The letter y is a variable.

b. The letters s and t both are variables.

EXAMPLE 1.2   Identify the constants and the variables in the algebraic expression

$$x + \frac{3}{2} + y + 1.4.$$

Solution

The fraction and decimal $\frac{3}{2}$ and 1.4 are numbers, so they are constants. The letters x and y are variables.

A **term** of an algebraic expression consists of products and quotients of constants and variables. The constants and variables within a term may be combined by multiplication and division operation symbols.

Four kinds of multiplication symbols are used in algebra. The ''times'' symbol, as in

$$2 \times 3$$

is rarely used because the symbol $\times$ is too easily confused with the letter x.

Multiplication is commonly indicated by a number and letter, or two or more letters, written next to each other. Thus,

$2x$ means the product of 2 and x.

xy means the product of x and y.

abc means the product of a, b, and c.

Of course, we cannot use this symbol for two numbers because 23 could then mean either the product of 2 and 3 or the number twenty-three.

A common multiplication symbol for two numbers, or sometimes a number and a letter, is a pair of parentheses. Thus,

$(2)(3)$ means the product of 2 and 3.

It is not necessary to have parentheses on both numbers:

$3(8)$ means the product of 3 and 8.

$t(16)$ means the product of t and 16.

A dot is also used to indicate multiplication. However, the dot is not often used as a multiplication symbol in algebra when decimals are involved because, for example, $2 \cdot 3$ is easily confused with the decimal 2.3.

A common division symbol in algebra is the *bar*. Thus, $\frac{a}{b}$ means the quotient of a and b. The symbol $\div$ is occasionally used.

EXAMPLE 1.3   Identify the constants and the variables in the algebraic expression

$$3xy + \frac{1}{2}z + 5.3t.$$

Solution   The constants are 3, $\frac{1}{2}$, and 5.3, and the variables are x, y, z, and t.

EXAMPLE 1.4   Identify the constants and the variables in the algebraic expression

$$rst(1.6).$$

Solution   The constant is the number 1.6, and the variables are the letters r, s, and t.

Terms of an algebraic expression are connected by addition and subtraction.

EXAMPLE 1.5   Determine the number of terms in each of these algebraic expressions:

a. $rst(1.6)$ b. $3xy + \frac{1}{2}z + 5.3t$

Solutions   a. The expression has one term since it contains only products, with no additions or subtractions.

b. The expression has three terms connected by additions: $3xy$, $\frac{1}{2}z$, and $5.3t$.

When a term of an algebraic expression contains both variables and a constant, the constant is called the **numerical coefficient** of the variables.

EXAMPLE 1.6 Identify the numerical coefficient of the variables in each of these terms:

$$\text{a. } 2xy \qquad \text{b. } z(2.3) \qquad \text{c. } \frac{1}{2}ax$$

Solutions

a. The numerical coefficient is 2.
b. The numerical coefficient is 2.3.
c. The numerical coefficient is $\frac{1}{2}$.

Observe that a quotient such as $\frac{x}{2}$ is the same as the product $\frac{1}{2}x$, since a number divided by 2 is the same as one-half of the number. Thus, a term such as

$$\frac{x}{2}$$

may be said to have the numerical coefficient of $\frac{1}{2}$.

EXAMPLE 1.7 Identify the numerical coefficient of x, the numerical coefficient of y, and the numerical coefficient of z in the algebraic expression

$$\frac{x}{4} + \frac{5y}{2} + (6.5)z + 6.$$

Solution

Since $\frac{x}{4}$ is the same as $\frac{1}{4}x$, the numerical coefficient of x is $\frac{1}{4}$. Similarly, the numerical coefficient of y is $\frac{5}{2}$. The numerical coefficient of z is 6.5. The term consisting of 6 alone contains no variables. Such a term is called a **constant term.**

A term such as xy appears to have no numerical coefficient. However, we may say that this term has the numerical coefficient 1 since $1(xy) = xy$.

EXAMPLE 1.8 Identify the numerical coefficient of x and the numerical coefficient of y in the algebraic expression

$$2x + y.$$

Solution   The numerical coefficient of x is 2. Since $1(y) = y$, the numerical coefficient of y is 1.

Exercise 1.1

Identify the constants and the variables in each of these algebraic expressions:

1. $x + 3$
2. $25 + x$
3. $u + v$
4. $y + z$

5. $t + u + \dfrac{5}{3}$
6. $\dfrac{1}{2} + x + y + z$
7. $\dfrac{3}{4}u + \dfrac{2}{5}v$
8. $10.2x + 5.4y$

9. $\dfrac{1}{2}xy$
10. $\dfrac{2}{3}pq$
11. $xyz(9.3)$
12. $(rs)(10.2)$

13. $st + (2)(3)$
14. $st + uv$

15–28. Determine the number of terms in each of the algebraic expressions in Exercises 1–14.

Identify the numerical coefficient in each of these terms:

29. $3st$

30. $4.4uv$

31. $\dfrac{1}{3}t$

32. $6.4xy$

33. $ab\left(\dfrac{1}{4}\right)$

34. $bcz(5.6)$

35. $\dfrac{x}{3}$

36. $\dfrac{2st}{3}$

Identify the numerical coefficient of x and the numerical coefficient of y in each of these algebraic expressions:

37. $\dfrac{x}{2} + y(1.3) + \dfrac{1}{3}$

38. $\dfrac{2y}{3} + 12.4x + 4.25$

39. $y + \dfrac{x}{4}$

40. $x + y$

1.2   Evaluating Expressions

In Section 1.1 we described an algebraic expression as a combination of variables, constants, and operation symbols that represents a number. We also described a variable in an algebraic expression as representing various numerical values. If we choose numerical values for the variables in an algebraic expression, we can calculate a numerical value for the expression. This process is called **evaluating** the algebraic expression.

 EXAMPLE 1.9   Evaluate the algebraic expression

$$5x + 3y + 6$$

if $x = 2$ and $y = 7$.

Solution   We start with the expression

$$5x + 3y + 6$$

and put 2 in place of x and 7 in place of y:

$$
\begin{aligned}
5x + 3y + 6 &= 5(2) + 3(7) + 6 \\
&= 10 + 21 + 6 \\
&= 37.
\end{aligned}
$$

Before we evaluate expressions involving fractions, we will review methods from arithmetic for multiplying and adding fractions.

Recall from arithmetic that, for a fraction $\dfrac{a}{b}$ where a and b are whole numbers with $b \neq 0$, a is called the **numerator** and b is called the **denominator.** To multiply two fractions, we simply multiply the numerators and multiply the denominators. Thus, for example,

$$\frac{1}{5} \cdot \frac{3}{8} = \frac{(1)(3)}{(5)(8)} = \frac{3}{40}.$$

If possible, we reduce the product by dividing any **common factors** from the numerator and the denominator:

$$\frac{1}{6} \cdot \frac{3}{8} = \frac{(1)(3)}{(6)(8)} = \frac{3}{48} = \frac{(1)(\cancel{3})}{(16)(\cancel{3})} = \frac{1}{16}.$$

You may divide out any common factors from the numerator and denominator before multiplying. For the preceding example,

$$\frac{1}{6}\cdot\frac{3}{8}=\frac{1}{(2)(\not3)}\cdot\frac{\not3}{8}=\frac{1}{(2)(8)}=\frac{1}{16}.$$

To multiply a whole number by a fraction, we may think of the whole number as a fraction with denominator 1. Thus,

$$1=\frac{1}{1}$$
$$2=\frac{2}{1}$$
$$3=\frac{3}{1}$$

and so on. To multiply

$$3\left(\frac{5}{24}\right),$$

we write

$$\frac{3}{1}\cdot\frac{5}{24}=\frac{\not3}{1}\cdot\frac{5}{(\not3)(8)}$$
$$=\frac{5}{(1)(8)}$$
$$=\frac{5}{8}.$$

Recall that to add two fractions, we must find a **common denominator.** For example, to add

$$\frac{3}{5}+\frac{2}{3},$$

we use the common denominator 15:

$$\frac{3}{5}+\frac{2}{3}=\frac{3(3)}{5(3)}+\frac{2(5)}{3(5)}$$
$$=\frac{9}{15}+\frac{10}{15}$$
$$=\frac{9+10}{15}$$
$$=\frac{19}{15}.$$

To add two fractions such as

$$\frac{1}{2}+\frac{3}{8},$$

we may use the common denominator $(2)(8) = 16$. However, the **least common denominator** is 8. Observe that 2 and 8 are both factors of 8, and there is no smaller number than 8 that has both 2 and 8 as factors. Thus, we use 8 as our common denominator:

$$\frac{1}{2} + \frac{3}{8} = \frac{1(4)}{2(4)} + \frac{3}{8}$$

$$= \frac{4}{8} + \frac{3}{8}$$

$$= \frac{4 + 3}{8}$$

$$= \frac{7}{8}.$$

If one of the numbers to be added is a whole number, again we write the whole number as a fraction, using the denominator 1. For example,

$$3 + \frac{5}{8} = \frac{3}{1} + \frac{5}{8}$$

$$= \frac{3(8)}{1(8)} + \frac{5}{8}$$

$$= \frac{24}{8} + \frac{5}{8}$$

$$= \frac{24 + 5}{8}$$

$$= \frac{29}{8}.$$

Now, we may evaluate algebraic expressions where the numerical coefficients or the values of the variables, or both, are fractions.

EXAMPLE 1.10   Evaluate the algebraic expression

$$\frac{1}{2}x + \frac{1}{3}y + 6$$

if $x = \frac{1}{5}$ and $y = 3$.

Solution   We put $\frac{1}{5}$ in place of x and 3 in place of y to obtain

$$\frac{1}{2}x + \frac{1}{3}y + 6 = \frac{1}{2}\left(\frac{1}{5}\right) + \frac{1}{3}(3) + 6$$

$$= \frac{1}{2} \cdot \frac{1}{5} + \frac{1}{3} \cdot \frac{3}{1} + \frac{6}{1}$$

$$= \frac{1}{(2)(5)} + \frac{(1)\cancel{(3)}}{\cancel{(3)}(1)} + \frac{6}{1}$$

$$= \frac{1}{10} + \frac{1}{1} + \frac{6}{1}.$$

The least common denominator is 10, thus

$$
\begin{aligned}
\frac{1}{10} + \frac{1}{1} + \frac{6}{1} &= \frac{1}{10} + \frac{1(10)}{1(10)} + \frac{6(10)}{1(10)} \\
&= \frac{1}{10} + \frac{10}{10} + \frac{60}{10} \\
&= \frac{1 + 10 + 60}{10} \\
&= \frac{71}{10}.
\end{aligned}
$$

EXAMPLE 1.11   Evaluate the algebraic expression

$$
\frac{st}{2} + u
$$

if $s = \frac{1}{3}$, $t = \frac{2}{5}$, and $u = \frac{3}{10}$.

Solution   Recall that we may think of

$$
\frac{st}{2} + u
$$

as

$$
\frac{1}{2}st + u.
$$

Then, replacing s by $\frac{1}{3}$, t by $\frac{2}{5}$, and u by $\frac{3}{10}$, we have

$$
\begin{aligned}
\frac{1}{2}st + u &= \frac{1}{2}\left(\frac{1}{3}\right)\left(\frac{2}{5}\right) + \frac{3}{10} \\
&= \frac{(1)(1)(2)}{(2)(3)(5)} + \frac{3}{10} \\
&= \frac{1}{15} + \frac{3}{10}.
\end{aligned}
$$

The least common denominator is 30:

$$
\begin{aligned}
\frac{1}{15} + \frac{3}{10} &= \frac{1(2)}{15(2)} + \frac{3(3)}{10(3)} \\
&= \frac{2}{30} + \frac{9}{30} \\
&= \frac{11}{30}.
\end{aligned}
$$

To evaluate algebraic expressions involving decimals, we recall the arithmetic of decimal numbers. To add or subtract decimals, we align the decimal points and add or subtract as usual. Thus, for example, to add 5.7 and 13.9, we write

$$5.7$$
$$\underline{13.9}$$
$$19.6$$

To subtract, for example, 1.25 from 2.2, we must supply a zero at the end of the decimal 2.2 and "borrow":

$$2.20$$
$$\underline{-\ 1.25}$$
$$0.95$$

To multiply two decimals, we use long multiplication, and then mark off the total number of decimal places in the factors. For example, to multiply 4.39 by 3.5, we mark off a total of three decimal places:

$$4.39$$
$$\underline{\quad 3.5}$$
$$2195$$
$$\underline{1317\quad}$$
$$15.365$$

Now, we may evaluate algebraic expressions involving decimal numbers.

EXAMPLE 1.12   Evaluate the algebraic expression

$$3x + 2y + 5$$

if $x = 2.8$ and $y = 6.3$.

Solution   Replacing x by 2.8 and y by 6.3,

$$
\begin{aligned}
3x + 2y + 5 &= 3(2.8) + 2(6.3) + 5 \\
&= 8.4 + 12.6 + 5 \\
&= 26.
\end{aligned}
$$

EXAMPLE 1.13   Evaluate the algebraic expression

$$3.2p + 10.4q$$

if $p = 9.5$ and $q = 5.5$.

Solution   Replacing p by 9.5 and q by 5.5,

$$
\begin{aligned}
3.2p + 10.4q &= 3.2(9.5) + 10.4(5.5) \\
&= 30.4 + 57.2 \\
&= 87.6.
\end{aligned}
$$

An algebraic expression may involve both fractions and decimal numbers. When we evaluate such an expression, we may have to round off to a given number of decimal places.

EXAMPLE 1.14   Evaluate the algebraic expression

$$\frac{1}{2}u + \frac{1}{3}v + 4.2w$$

if $u = 3.5$, $v = 12.5$, and $w = \frac{3}{4}$, to two decimal places.

Solution

$$\frac{1}{2}u + \frac{1}{3}v + 4.2w = \frac{1}{2}(3.5) + \frac{1}{3}(12.5) + 4.2\left(\frac{3}{4}\right).$$

To evaluate the first term, we think of $\frac{1}{2}$ as a division by 2:

$$\frac{1}{2}(3.5) = \frac{3.5}{2} = 1.75.$$

The division by 3 in the second term does not terminate:

$$\frac{1}{3}(12.5) = \frac{12.5}{3} = 4.1666 \ldots \approx 4.17.$$

To evaluate the third term, we multiply by 3 and divide by 4:

$$4.2\left(\frac{3}{4}\right) = \frac{4.2(3)}{4} = \frac{12.6}{4} = 3.15.$$

The value of the expression is

$$1.75 + 4.17 + 3.15 = 9.07$$

to two decimal places.

Throughout this book we will use conventional rounding: If the digit following the given decimal place is less than 5, drop it and all following digits. If the digit following the given decimal place is 5 or more, increase the digit in the given place by 1 and drop all following digits.

Exercise 1.2

Evaluate each expression for the given values of the variables:

1. $5x + 3y$ if $x = 9$ and $y = 10$

2. $6s + 8t + 10$ if $s = 2$ and $t = 4$

3. $\frac{1}{2}u + \frac{1}{3}v + 4$ if $u = 8$ and $v = \frac{1}{3}$

4. $\frac{2}{3}p + \frac{4}{3}q + \frac{2}{9}$ if $p = \frac{2}{3}$ and $q = 2$

5. $\frac{3}{5}t + \frac{3}{4}u$ if $t = \frac{3}{4}$ and $u = \frac{2}{5}$

6. $\frac{1}{2}y + \frac{1}{4}z$ if $y = \frac{1}{3}$ and $z = \frac{1}{2}$

7. $3y + 9z$ if $y = \frac{2}{3}$ and $z = \frac{4}{3}$

8. $8r + 9s + 1$ if $r = \frac{5}{4}$ and $s = \frac{5}{3}$

9. $\frac{x}{2} + \frac{y}{3}$ if $x = \frac{4}{3}$ and $y = \frac{4}{5}$

10. $\frac{xy}{5} + \frac{2z}{3}$ if $x = \frac{1}{2}$, $y = \frac{3}{4}$, and $z = \frac{2}{5}$

11. $3t + 2u$ if $t = 9.8$ and $u = 8.8$

12. $3x + 4y + z$ if $x = 2.9$, $y = 3.4$, and $z = 2.3$

13. $4.5s + 9.5t$ if $s = 10.2$ and $t = 5.2$

14. $3.3p + 1.56q$ if $p = 0.39$ and $q = 1.2$

15. $\frac{1}{2}r + 2.2s + \frac{1}{3}t$ if $r = 3.5$, $s = 1.25$, and $t = 3.5$, to two decimal places

16. $\frac{u}{4} + \frac{v}{3} + \frac{2w}{3}$ if $u = 12.6$, $v = 5.8$, and $w = 9.4$, to two decimal places

1.3 The Associative and Commutative Properties

There are two basic properties we use in dealing with numbers in arithmetic. First, if we are adding three numbers, we usually add the first two and then the third. Thus, to add

$$8 + 3 + 9,$$

we add

$$8 + 3 = 11$$

and

$$11 + 9 = 20;$$

therefore,

$$8 + 3 + 9 = 20.$$

To show we added the first two numbers and then the third, we use parentheses as a grouping symbol:

$$(8 + 3) + 9 = 11 + 9 = 20.$$

(Recall that parentheses have been previously used as a multiplication symbol, which is an entirely different use.) We may also add the second two numbers, and then add the first to the result:

$$3 + 9 = 12$$

and

$$8 + 12 = 20.$$

To show we added the second two numbers first, we use parentheses again as a grouping symbol:

$$8 + (3 + 9) = 8 + 12 = 20.$$

Observe that the result in either case is the same:

$$(8 + 3) + 9 = 11 + 9 = 20$$

and

$$8 + (3 + 9) = 8 + 12 = 20.$$

Therefore, the two expressions are equal:

$$(8 + 3) + 9 = 8 + (3 + 9).$$

In general, we may group the addition of three numbers in either way without changing the result. Since the statement is true for any three numbers, we write an algebraic statement using variables for the numbers. The statement is called the **associative property for addition.**

Associative Property for Addition: For any three numbers x, y, and z,

$$(x + y) + z = x + (y + z).$$

A similar property is true for multiplication. Consider the product

$$(4)(3)(5).$$

We may multiply the first two numbers, and then multiply that product by the third number:

$$(4)(3) = 12$$

and

$$(12)(5) = 60.$$

To show we multiplied the first two numbers and then the third, we may use grouping brackets and multiplication parentheses:

$$[(4)(3)](5) = (12)(5) = 60.$$

Alternatively, we may multiply the second two numbers, and then multiply the first by that product:

$$(3)(5) = 15$$

and

$$(4)(15) = 60.$$

Using grouping brackets to show we multiplied the second two numbers,

$$(4)[(3)(5)] = (4)(15) = 60.$$

The two expressions are numerically equal; therefore,

$$[(4)(3)](5) = (4)[(3)(5)].$$

We write an algebraic statement for this equality using variables. Observe that we do not need a multiplication symbol, since we may simply write the letters next to one another. The parentheses are grouping parentheses, not multiplication parentheses. The statement is called the **associative property for multiplication.**

Associative Property for Multiplication: For any three numbers x, y, and z,

$$(xy)z = x(yz).$$

In using the associative properties, we do not change the order in which the numbers are written. We use grouping parentheses to show which two of the three numbers are associated. However, there are properties by which we may change the order in which numbers are added or multiplied. We know that, for example,

$$5 + 8 = 13$$

and also

$$8 + 5 = 13.$$

Therefore,

$$5 + 8 = 8 + 5.$$

The algebraic statement for this equality using variables is called the **commutative property for addition.**

Commutative Property for Addition: For any two numbers x and y,

$$x + y = y + x.$$

When we write algebraic expressions involving sums of variables, we usually write the variables in alphabetical order. Thus, since $s + t = t + s$, we usually write $s + t$ for either $s + t$ or $t + s$. In writing the sum of a variable and a numerical constant, we usually write the letter first. Thus we would write $s + 16$ rather than $16 + s$, and $a + 2$ rather than $2 + a$.

For multiplication, observe that, for example,

$$(4)(9) = 36$$

and

$$(9)(4) = 36.$$

Therefore,

$$(4)(9) = (9)(4).$$

In writing the algebraic statement for this equality using variables, we simply write the letters next to one another. The statement is called the **commutative property for multiplication.**

Commutative Property for Multiplication: For any two numbers x and y,

$$xy = yx.$$

Again, for products, we usually write variables in alphabetical order. Since $st = ts$, we write st for either st or ts. For products, numerical coefficients usually are written first. Thus, we write $2x$ rather than $x2$ or $x(2)$, and $\frac{1}{3}a$ rather than $a\frac{1}{3}$ or $a\left(\frac{1}{3}\right)$.

The associative and commutative properties are often used together. For example, to add

$$4 + 8 + 6$$

it is common to say

$$4 + 6 = 10$$

and

$$10 + 8 = 18.$$

We have used the commutative property to say

$$4 + (8 + 6) = 4 + (6 + 8),$$

and the associative property to say

$$4 + (6 + 8) = (4 + 6) + 8.$$

EXAMPLE 1.15   State whether each of these algebraic statements is an example of the associative property or the commutative property:

a. $a(6) = 6a$ b. $a + (b + c) = a + (c + b)$ c. $(a + b) + 6 = a + (b + 6)$

Solutions

a. Although parentheses appear, these are multiplication parentheses. The order is changed, and so the statement is an example of the commutative property.
b. Again, although parentheses appear and they are in fact grouping parentheses, the grouping of the letters is not changed. The order is changed; therefore, the statement is an example of the commutative property.
c. The grouping is changed by the grouping parentheses, and the order is not changed. The statement is an example of the associative property.

EXAMPLE 1.16   State whether this algebraic statement is an example of the associative property or the commutative property:

$$(bc)a = (ab)c.$$

Solution   Both the associative and commutative properties are used. Using the commutative property, we may change the order of the first expression to

$$(bc)a = a(bc).$$

Then, using the associative property,

$$a(bc) = (ab)c.$$

☐ **Exercise 1.3**

State whether each of these algebraic statements is an example of the associative property or the commutative property or both:

1. $(3)(5) = (5)(3)$

2. $\left(\dfrac{1}{2} + \dfrac{3}{4}\right) + \dfrac{5}{6} = \dfrac{1}{2} + \left(\dfrac{3}{4} + \dfrac{5}{6}\right)$ —

3. $(pq)r = p(qr)$

4. $n + m = m + n$ —

5. $z\left(\dfrac{1}{2}\right) = \dfrac{1}{2}z$

6. $\left(s\dfrac{2}{3}\right)t = \left(\dfrac{2}{3}s\right)t$

7. $a\left(b\dfrac{1}{2}\right) = \left(ab\right)\dfrac{1}{2}$

8. $\dfrac{1}{2}(uv) = (uv)\dfrac{1}{2}$

9. $2 + (x + y) = (x + y) + 2$

10. $(c + d) + 3.5 = c + (d + 3.5)$

11. $2 + (7 + 8) = (2 + 8) + 7$

12. $(a3)z = 3(az)$

☐ **1.4** # The Distributive Property

A basic property of arithmetic relates addition and multiplication. Suppose we add two numbers and then multiply the result by another number. For example,

$$4 + 7 = 11$$

and

$$2(11) = 22.$$

We may write

$$2(4 + 7) = 2(11) = 22,$$

where the parentheses serve as both grouping and multiplication parentheses.

Now, suppose we multiply each of the original numbers by 2 first, and then add the products.

$$2(4) = 8$$
$$2(7) = 14$$

and

$$8 + 14 = 22.$$

We may write

$$2(4) + 2(7) = 8 + 14 = 22.$$

Here the parentheses are only multiplication parentheses.

The two expressions are numerically equal; therefore,

$$2(4 + 7) = 2(4) + 2(7).$$

We can write this statement as an algebraic statement, using variables for the numbers. The statement is called the **distributive property for multiplication over addition.**

Distributive Property for Multiplication over Addition: For any three numbers x, y, and z,

$$x(y + z) = xy + xz.$$

Recall that the parentheses in the second part of the numerical statement were multiplication parentheses. Since the algebraic statement uses letters, multiplication parentheses are not necessary. It is understood that, in the second part, we do the multiplications first and then add the products. Thus, there are also no grouping parentheses in the second part.

We may use the distributive property to remove parentheses when an expression is multiplied by a constant.

EXAMPLE 1.17   Use the distributive property to remove the parentheses in the algebraic expression

$$2(x + y).$$

Solution   Using the distributive property,

$$2(x + y) = 2x + 2y.$$

EXAMPLE 1.18   Use the distributive property to remove the parentheses in the algebraic expression

$$3(4a + 5b).$$

Solution   Using the distributive property,

$$3(4a + 5b) = 3(4a) + 3(5b).$$

We still have multiplication parentheses. However, using the associative property,

$$3(4a) = (3)(4)a = 12a,$$

and

$$3(5b) = (3)(5)b = 15b.$$

Therefore,

$$3(4a + 5b) = 3(4a) + 3(5b)$$
$$= 12a + 15b.$$

EXAMPLE 1.19   Use the distributive property to remove the parentheses in the algebraic expression

$$4\left(\frac{1}{2}p + \frac{2}{3}q\right) + \frac{2}{3}\left(3r + \frac{5}{6}s\right).$$

Solution   We apply the distributive property separately to each set of parentheses:

$$4\left(\frac{1}{2}p + \frac{2}{3}q\right) + \frac{2}{3}\left(3r + \frac{5}{6}s\right) = 4\left(\frac{1}{2}p\right) + 4\left(\frac{2}{3}q\right) + \frac{2}{3}\left(3r\right) + \frac{2}{3}\left(\frac{5}{6}s\right).$$

Then, using the associative property within each term, and recalling our discussion of multiplication of fractions,

$$4\left(\frac{1}{2}p\right) + 4\left(\frac{2}{3}q\right) + \frac{2}{3}\left(3r\right) + \frac{2}{3}\left(\frac{5}{6}s\right) = \left(\frac{4}{1} \cdot \frac{1}{2}\right)p + \left(\frac{4}{1} \cdot \frac{2}{3}\right)q + \left(\frac{2}{3} \cdot \frac{3}{1}\right)r + \left(\frac{2}{3} \cdot \frac{5}{6}\right)s$$

$$= 2p + \frac{8}{3}q + 2r + \frac{5}{9}s.$$

If the algebraic expression in the parentheses has more than two terms, we simply multiply each term, however many, by the constant. This property is called the **extended distributive property.** However, we will refer to it also as the distributive property.

EXAMPLE 1.20   Use the distributive property to remove the parentheses in the algebraic expression

$$\frac{3}{8}\left(\frac{4}{9}u + \frac{5}{12}v + \frac{8}{15}w\right).$$

Solution   Using the extended distributive property,

$$\frac{3}{8}\left(\frac{4}{9}u + \frac{5}{12}v + \frac{8}{15}w\right) = \frac{3}{8}\left(\frac{4}{9}u\right) + \frac{3}{8}\left(\frac{5}{12}v\right) + \frac{3}{8}\left(\frac{8}{15}w\right)$$

$$= \left(\frac{3}{8} \cdot \frac{4}{9}\right)u + \left(\frac{3}{8} \cdot \frac{5}{12}\right)v + \left(\frac{3}{8} \cdot \frac{8}{15}\right)w$$

$$= \frac{1}{6}u + \frac{5}{32}v + \frac{1}{5}w.$$

Exercise 1.4

Use the distributive property to remove the parentheses in each of these algebraic expressions:

1. $3(p + q)$

2. $\frac{1}{2}(a + b)$

3. $a(x + y)$

4. $2x(s + t)$

5. $5(3a + 2b)$

6. $8(12y + 14z)$

7. $\frac{1}{4}(2m + 8n)$

8. $6\left(\frac{2}{3}u + \frac{5}{8}v\right)$

9. $2\left(\dfrac{3}{2}a + \dfrac{5}{4}b\right) + \dfrac{3}{2}\left(4c + \dfrac{5}{6}d\right)$

10. $\dfrac{2}{5}\left(\dfrac{9}{10}w + \dfrac{3}{4}x\right) + \dfrac{3}{4}\left(\dfrac{2}{9}y + \dfrac{4}{15}z\right)$

11. $6\left(\dfrac{2}{3}p + \dfrac{3}{4}q + \dfrac{5}{6}r\right)$

12. $\dfrac{8}{15}(10t + 5u + 9v + 3w)$

13. $2.5(2x + 4y)$

14. $5(3.2r + 5.6s)$

15. $1.2(3.3x + 2.3y) + 2.2(4.3z + 1.3u)$

16. $3.92(4.5p + 6.7q + 9.3r)$

Self-test

1. Identify the constants and the variables in this algebraic expression:

   $3x + \dfrac{1}{2}y + 2.5$

   1. constants _____ 2 5 _____

      variables _____ x, y _____

2. Identify the numerical coefficient of x and the numerical coefficient of y in this algebraic expression:

   $\dfrac{2}{3}x + y(3.3) + 10.2$

   2. coefficient of x _____ 2/3, ___ _____

      coefficient of y _____ 3.3 _____

3. Evaluate the algebraic expression

   $\dfrac{1}{4}x + \dfrac{y}{3} + 4z$

   if $x = \frac{2}{3}$, $y = \frac{4}{3}$, and $z = \frac{1}{2}$.

   3. _____ $\dfrac{47}{18}$ _____

4. Use the distributive property to remove the parentheses in the algebraic expression

   $\dfrac{1}{2}(2x + 5y) + \dfrac{1}{3}(6u + 4v)$.

   4. _____

5. State whether each of these algebraic statements is an example of the associative property or the commutative property:

   a. $(s + t) + 3 = s + (t + 3)$

   b. $y(2) = 2y$

   5a. _____

   5b. _____

UNIT

2 Basic Equations

INTRODUCTION

At the beginning of the preceding unit, you read that the word *algebra* derives from a word meaning "to restore." This meaning refers to the process of equation solving. Equation solving was the original purpose of algebra. In this unit you will learn what an equation is, and what is meant by a solution of an equation. You will learn the four basic methods for solving equations. Finally, you will learn several applications of one type of equation.

OBJECTIVES

When you have finished this unit you should be able to:

1. Determine whether or not a given number is a solution of a given equation.
2. Solve equations of the form $x + a = b$ and $x - a = b$.
3. Solve equations of the form $ax = b$ and $\frac{x}{a} = b$.
4. Use an equation of the form $ax = b$ to solve percent, uniform motion, and unit price problems.

| 2.1 | ## Solutions of Equations

In sixteenth-century Italy, solving equations of higher orders was not only a primary mathematical activity but also a pastime. Mathematicians challenged one another to contests in equation solving. As much gamblers as they were mathematicians, some had few scruples about where and how they got their solutions.

An **equation** consists of two algebraic expressions with an equal sign between them. These are some examples of equations:

$$x + 3 = 10, \quad \frac{1}{2}x - \frac{3}{2} = \frac{2}{3}, \quad 4x - (2x - 1) = 3(x + 2), \quad 3x + 2y - 6 = 0.$$

An equation is **in one variable** if it contains only one variable. These are examples of equations in one variable:

$$x + 3 = 10, \quad \frac{1}{2}x - \frac{3}{2} = \frac{2}{3}, \quad 4x - (2x - 1) = 3(x + 2).$$

Observe that the one variable may occur more than once.

The algebraic expressions on the two sides of the equal sign are called the **sides** of the equation. The expression to the left of the equal sign is the **left-hand side** of the equation. The expression to the right of the equal sign is the **right-hand side** of the equation.

Suppose we evaluate the algebraic expressions on each side of an equation in one variable for a specific numerical value of the variable. We say we have substituted the number for the variable. If substituting a number for the variable causes the two sides of the equation to be numerically equal, then the number is a **solution** of the equation.

EXAMPLE 2.1 Determine whether $x = 3$ is a solution of the equation $x + 10 = 13$.

Solution We substitute 3 for x in the equation. When x is replaced by 3 we have

$$x + 10 = 13$$
$$3 + 10 \stackrel{?}{=} 13$$
$$13 = 13.$$

The two sides are numerically equal; therefore, $x = 3$ is a solution of the equation $x + 10 = 13$.

EXAMPLE 2.2 Determine whether $t = \frac{3}{4}$ is a solution of the equation $2t - \frac{1}{2} = \frac{3}{2}$.

Solution We substitute $\frac{3}{4}$ for t in the equation:

$$2t - \frac{1}{2} = \frac{3}{2}$$
$$2\left(\frac{3}{4}\right) - \frac{1}{2} \stackrel{?}{=} \frac{3}{2}$$
$$\frac{3}{2} - \frac{1}{2} \stackrel{?}{=} \frac{3}{2}$$
$$\frac{2}{2} \stackrel{?}{=} \frac{3}{2}$$
$$1 \neq \frac{3}{2}.$$

The symbol $\neq$ means "not equal to": 1 is not equal to $\frac{3}{2}$. Thus, $t = \frac{3}{4}$ is not a solution of the equation.

EXAMPLE 2.3 Determine whether $u = \frac{2}{3}$ is a solution of the equation $9u - 2 = 3u + 1$.

Solution We substitute $\frac{2}{3}$ for u, which means we must replace u by $\frac{2}{3}$ whenever u appears:

$$9u - 2 = 3u + 1$$
$$9\left(\frac{2}{3}\right) - 2 \stackrel{?}{=} 3\left(\frac{2}{3}\right) + 1$$
$$3(2) - 2 \stackrel{?}{=} 2 + 1$$
$$6 - 2 \stackrel{?}{=} 3$$
$$4 \neq 3.$$

Therefore, $u = \frac{2}{3}$ is not a solution of the equation.

EXAMPLE 2.4 Determine whether $s = 2.3$ is a solution of the equation $4s + 1.3 = 6s - 3.3$.

Solution   Substituting 2.3 for s whenever s appears,

$$4s + 1.3 = 6s - 3.3$$
$$4(2.3) + 1.3 \stackrel{?}{=} 6(2.3) - 3.3$$
$$9.2 + 1.3 \stackrel{?}{=} 13.8 - 3.3$$
$$10.5 = 10.5.$$

Therefore, $s = 2.3$ is a solution of the equation.

Exercise 2.1

Determine whether the given number is a solution of the equation:

1. $x + 16 = 20; x = 4$

2. $y - 8 = 5; y = 14$

3. $2r + 7 = 8; r = \dfrac{1}{4}$

4. $4p - 9 = 1; p = \dfrac{5}{2}$

5. $5u - \dfrac{2}{3} = \dfrac{1}{6}; u = \dfrac{1}{6}$

6. $\dfrac{1}{4}v + \dfrac{3}{8} = \dfrac{1}{4}; v = \dfrac{1}{2}$

7. $9u - 3 = 3u + 1; u = \dfrac{2}{3}$

8. $4w + 3 = 2w + 6; w = \dfrac{3}{2}$

9. $4 - \dfrac{1}{2}s = 3 + \dfrac{3}{2}s; s = \dfrac{1}{2}$

10. $\dfrac{1}{3} + \dfrac{3}{2}t = \dfrac{2}{3} - \dfrac{1}{2}t; t = \dfrac{1}{3}$

11. $2z + 5 = 4z + 2; z = 1.5$

12. $3x - 9 = 24 - 3x; x = 5.5$

13. $6.3t + 2.2 = 4.3t + 4.4; t = 2.1$

14. $9.2s - 1.2 = 3s + 14.3; s = 2.5$

15. $\dfrac{1}{3}y + \dfrac{2}{3}(y + 0.25) = \dfrac{2}{3}y + \dfrac{1}{3}; y = 0.75$

16. $0.75z + 0.5\left(z + \dfrac{2}{3}\right) = 0.5z + \dfrac{1}{3}(z + 1.5); z = 0.4$

2.2 Solving by Subtraction and Addition

Suppose we are given a linear equation in one variable. To find a solution of the equation, we try to isolate the variable on one side of the equation, with a single number on the other side. That number is then the solution. This process is called **solving an equation.**

To solve an equation, we use a series of **equivalent equations.** Equivalent equations have the same solution as the given equation. For example,

$$4x - 5 = 9$$

and

$$4x = 14$$

are equivalent equations. You should check that each equation has the solution $x = \frac{7}{2}$.

In deriving equivalent equations, we must always adjust one side for every change made on the other side, to be sure the equality is maintained. This adjustment is what was meant by "restoring" balance in the original meaning of the word *algebra*. There are four basic methods for deriving equivalent equations.

Method 1: An equivalent equation can be derived from an equation by subtracting the same number from both sides.

Before considering examples of Method 1, we review an important property of numbers. If we add and subtract the same number, the result is a net change of zero. For example, if you earn $5 (adding) and then spend $5 (subtracting), you have the same amount that you started with. If you have x dollars and earn $5, you have $x + 5$ dollars. If you then spend $5, you have

$$(x + 5) - 5 = x + (5 - 5)$$
$$= x + 0$$
$$= x,$$

or x dollars.

Now, suppose we have an equation such as

$$x + 5 = 7.$$

Using Method 1, we may derive an equivalent equation by subtracting the same number from both sides. If we choose to subtract 5, we will isolate x:

$$
\begin{array}{rr}
x + 5 = & 7 \\
- 5 & - 5 \\
\hline
x + 0 = & 2. \\
x = & 2.
\end{array}
$$

By subtracting 5 from both sides of the equation $x + 5 = 7$, we have derived the equivalent equation $x = 2$. We have also found the solution $x = 2$ of the original equation. You can check this solution by substituting $x = 2$ into the original equation.

EXAMPLE 2.5 Solve the equation $x + 6 = 21$ and check the solution.

Solution   We subtract 6 from each side:

$$
\begin{array}{rr}
x + 6 = & 21 \\
- 6 & - 6 \\
\hline
x + 0 = & 15 \\
x = & 15.
\end{array}
$$

To check, we substitute $x = 15$ in the original equation:

$$x + 6 = 21$$
$$15 + 6 \stackrel{?}{=} 21$$
$$21 = 21.$$

The two sides are numerically equal; therefore, $x = 15$ is the solution of the equation.

EXAMPLE 2.6   Solve the equation $u + 2.3 = 4.2$ and check the solution.

Solution   We must isolate the variable u. To do so, we derive an equivalent equation by subtracting 2.3 from each side:

$$
\begin{array}{rcr}
u + 2.3 = & & 4.2 \\
- 2.3 & & - 2.3 \\
\hline
u + \quad 0 = & & 1.9 \\
u = & & 1.9.
\end{array}
$$

You should check this solution by substituting $u = 1.9$ in the original equation.

EXAMPLE 2.7   Solve the equation $z + \frac{3}{4} = \frac{7}{8}$ and check the solution.

Solution   We subtract $\frac{3}{4}$ from each side:

$$
\begin{array}{rcl}
z + \dfrac{3}{4} = & & \dfrac{7}{8} \\[2mm]
- \dfrac{3}{4} & & - \dfrac{3}{4} \\[2mm]
\hline
z + 0 = & & \dfrac{7}{8} - \dfrac{3}{4} \\[2mm]
z = & & \dfrac{7}{8} - \dfrac{3}{4}.
\end{array}
$$

To reduce this solution to a single number, we use a common denominator:

$$z = \frac{7}{8} - \frac{6}{8}$$
$$z = \frac{7 - 6}{8}$$
$$z = \frac{1}{8}.$$

We check by substituting $z = \frac{1}{8}$ in the original equation:

$$z + \frac{3}{4} = \frac{7}{8}$$
$$\frac{1}{8} + \frac{3}{4} \stackrel{?}{=} \frac{7}{8}$$

$$\frac{1}{8} + \frac{6}{8} \stackrel{?}{=} \frac{7}{8}$$

$$\frac{1 + 6}{8} \stackrel{?}{=} \frac{7}{8}$$

$$\frac{7}{8} = \frac{7}{8}.$$

Therefore, $z = \frac{1}{8}$ is the solution of the equation.

Method 2: An equivalent equation can be derived from an equation by adding the same number to both sides.

Method 2 depends on the fact that subtracting and then adding the same number results in a net change of zero.

EXAMPLE 2.8 Solve the equation $y - 6 = 11$ and check the solution.

Solution   We add 6 to each side:

$$\begin{array}{rcr} y - 6 = & 11 \\ + 6 & + 6 \\ \hline y + 0 = & 17 \\ y = & 17. \end{array}$$

To check, we substitute $y = 17$ in the original equation:

$$\begin{array}{rcl} y - 6 &=& 11 \\ 17 - 6 &\stackrel{?}{=}& 11 \\ 11 &=& 11. \end{array}$$

Therefore, $y = 17$ is the solution of the equation.

EXAMPLE 2.9 Solve the equation $t - 3.6 = 0.4$ and check the solution.

Solution   Adding 3.6 to each side,

$$\begin{array}{rcr} t - 3.6 = & 0.4 \\ + 3.6 & + 3.6 \\ \hline t + 0 = & 4.0 \\ t = & 4. \end{array}$$

You should check this solution by substituting $t = 4$ in the original equation.

EXAMPLE 2.10 Solve the equation $v - \frac{5}{6} = \frac{3}{8}$ and check the solution.

Solution   Adding $\frac{5}{6}$ to each side,

$$v - \frac{5}{6} = \frac{3}{8}$$

$$\underline{+\frac{5}{6} \qquad +\frac{5}{6}}$$

$$v + 0 = \frac{3}{8} + \frac{5}{6}$$

$$v = \frac{3}{8} + \frac{5}{6}$$

$$v = \frac{9}{24} + \frac{20}{24}$$

$$v = \frac{9 + 20}{24}$$

$$v = \frac{29}{24}.$$

To check,

$$v - \frac{5}{6} = \frac{3}{8}$$

$$\frac{29}{24} - \frac{5}{6} \stackrel{?}{=} \frac{3}{8}$$

$$\frac{29}{24} - \frac{20}{24} \stackrel{?}{=} \frac{3}{8}$$

$$\frac{29 - 20}{24} \stackrel{?}{=} \frac{3}{8}$$

$$\frac{9}{24} \stackrel{?}{=} \frac{3}{8}$$

$$\frac{3}{8} = \frac{3}{8}.$$

Therefore, $v = \frac{29}{24}$ is the solution of the equation.

Exercise 2.2

Solve and check:

1. $t + 10 = 30$ 2. $v + 13 = 25$ 3. $y + 4.9 = 8.1$ 4. $x + 0.12 = 0.2$

5. $x + \frac{1}{2} = \frac{3}{4}$ 6. $u + \frac{3}{5} = \frac{9}{10}$ 7. $s - 14 = 8$ 8. $u - 22 = 3$

9. $z - 2.3 = 3.8$ 10. $y - 31.5 = 3.5$ 11. $v - \frac{1}{3} = \frac{1}{5}$ 12. $y - \frac{3}{4} = \frac{4}{5}$

13. $z + \frac{1}{2} = \frac{17}{16}$ 14. $w - \frac{3}{5} = \frac{3}{25}$ 15. $y - \frac{7}{8} = \frac{11}{12}$ 16. $x + \frac{8}{9} = \frac{22}{15}$

17. $s - 4.43 = 6.79$ 18. $t + 0.95 = 1.85$ 19. $u + 3.6 = 3.6$ 20. $v - 2.5 = 2.5$

| 2.3 | **Solving by Division and Multiplication** |

In Section 2.2, we derived equivalent equations by adding or subtracting. There are two more methods for deriving equivalent equations.

> *Method 3:*   An equivalent equation can be derived from an equation by dividing both sides by the same number (where the number is not zero).

Observe that if we divide a number by itself, the result is 1. For example,

$$\frac{5x}{5} = \left(\frac{5}{5}\right)x$$
$$= (1)x$$
$$= x.$$

The number cannot be zero because division by zero is not defined, as we will explain in Unit 4.
    Now, suppose we have an equation such as

$$5x = 15.$$

Using Method 3, we may derive an equivalent equation by dividing both sides by the same number. If we choose to divide by 5, we will isolate x:

$$\frac{5x}{5} = \frac{15}{5}$$
$$1x = \frac{15}{5}$$
$$x = \frac{15}{5}$$
$$x = 3.$$

By dividing both sides of the equation $5x = 15$ by 5, we have derived the equivalent equation $x = 3$. Therefore, $x = 3$ is the solution of the original equation. You can check this solution by substituting $x = 3$ in the original equation.

EXAMPLE 2.11   Solve the equation $8x = 48$ and check the solution.

Solution   We divide each side by 8:

$$8x = 48$$
$$\frac{8x}{8} = \frac{48}{8}$$
$$1x = \frac{48}{8}$$
$$x = \frac{48}{8}$$
$$x = 6.$$

To check, we substitute $x = 6$ in the original equation:

$$8x = 48$$
$$8(6) \stackrel{?}{=} 48$$
$$48 = 48.$$

Therefore, $x = 6$ is the solution of the equation.

EXAMPLE 2.12   Solve the equation $12y = 16$ and check the solution.

Solution   We divide each side by 12:

$$12y = 16$$
$$\frac{12y}{12} = \frac{16}{12}$$
$$1y = \frac{16}{12}$$
$$y = \frac{16}{12}$$
$$y = \frac{4}{3}.$$

To check, we substitute $y = \frac{4}{3}$ in the original equation:

$$12y = 16$$
$$12\left(\frac{4}{3}\right) \stackrel{?}{=} 16$$
$$4(4) \stackrel{?}{=} 16$$
$$16 = 16.$$

Therefore, $y = \frac{4}{3}$ is the solution of the equation.

EXAMPLE 2.13   Solve the equation $0.25s = 1.5$ and check the solution.

Solution   We divide each side by 0.25:

$$0.25s = 1.5$$
$$\frac{0.25s}{0.25} = \frac{1.5}{0.25}$$
$$1s = \frac{1.5}{0.25}$$
$$s = \frac{1.5}{0.25}.$$

Recall from arithmetic that to divide decimals, we move each decimal point to the right the number of decimal places in the divisor. Thus,

$$s = \frac{150}{25}$$

$$s = 6.$$

To check, we substitute $s = 6$ in the original equation:

$$0.25s = 1.5$$

$$0.25(6) \stackrel{?}{=} 1.5$$

$$1.5 = 1.5.$$

Therefore, $s = 6$ is the solution of the equation.

EXAMPLE 2.14   Solve the equation $\frac{2}{3}t = \frac{8}{9}$ and check the solution.

Solution   We divide each side by $\frac{2}{3}$:

$$\frac{2}{3}t = \frac{8}{9}$$

$$\frac{\frac{2}{3}t}{\frac{2}{3}} = \frac{\frac{8}{9}}{\frac{2}{3}}$$

$$1t = \frac{\frac{8}{9}}{\frac{2}{3}}$$

$$t = \frac{\frac{8}{9}}{\frac{2}{3}}.$$

Recall that to divide two fractions, we multiply by the reciprocal of the divisor. (You may know this method as ''invert and multiply.'') The reciprocal of $\frac{2}{3}$ is $\frac{3}{2}$; therefore,

$$t = \frac{8}{9} \cdot \frac{3}{2}$$

$$t = \frac{4}{3}.$$

To check, we substitute $t = \frac{4}{3}$ in the original equation:

$$\frac{2}{3}t = \frac{8}{9}$$

$$\frac{2}{3}\left(\frac{4}{3}\right) \stackrel{?}{=} \frac{8}{9}$$

$$\frac{8}{9} = \frac{8}{9}.$$

Method 4: An equivalent equation can be derived from an equation by multiplying both sides by the same number (where the number is not zero).

EXAMPLE 2.15   Solve the equation $\frac{u}{7} = 3$ and check the solution.

Solution   We multiply each side by 7:

$$\frac{u}{7} = 3$$

$$7\left(\frac{u}{7}\right) = 7(3).$$

We observe that

$$7\left(\frac{u}{7}\right) = \left(\frac{7}{1} \cdot \frac{1}{7}\right)u$$

$$= (1)u$$

$$= u.$$

Therefore,

$$7\left(\frac{u}{7}\right) = 7(3)$$

$$1u = 7(3)$$

$$u = 7(3)$$

$$u = 21.$$

To check, we substitute $u = 21$ in the original equation:

$$\frac{u}{7} = 3$$

$$\frac{21}{7} \stackrel{?}{=} 3$$

$$3 = 3.$$

Therefore, $u = 21$ is the solution of the equation.

EXAMPLE 2.16   Solve the equation $\frac{s}{6} = \frac{4}{15}$ and check the solution.

Solution   Multiplying each side by 6,

$$\frac{s}{6} = \frac{4}{15}$$

$$6\left(\frac{s}{6}\right) = 6\left(\frac{4}{15}\right)$$

$$1s = 6\left(\frac{4}{15}\right)$$

$$s = 6\left(\frac{4}{15}\right)$$

$$s = \frac{8}{5}.$$

To check, we substitute $s = \frac{8}{5}$ in the original equation:

$$\frac{s}{6} = \frac{4}{15}$$

$$\frac{\frac{8}{5}}{6} \overset{?}{=} \frac{4}{15}$$

$$\frac{8}{5} \cdot \frac{1}{6} \overset{?}{=} \frac{4}{15}$$

$$\frac{4}{15} = \frac{4}{15}.$$

Therefore, $s = \frac{8}{5}$ is the solution of the equation.

In Example 2.14, we solved an equation involving fractions using Method 3. It is sometimes easier to solve such equations using Method 4.

EXAMPLE 2.17 Solve the equation $\frac{2}{3}t = \frac{8}{9}$.

Solution First, we multiply each side by 3:

$$3\left(\frac{2}{3}t\right) = 3\left(\frac{8}{9}\right)$$

$$1(2t) = 3\left(\frac{8}{9}\right)$$

$$2t = 3\left(\frac{8}{9}\right)$$

$$2t = \frac{8}{3}.$$

Now, we multiply each side by $\frac{1}{2}$, recalling that multiplication by $\frac{1}{2}$ is the same as division by 2:

$$\frac{1}{2}(2t) = \frac{1}{2}\left(\frac{8}{3}\right)$$

$$1t = \frac{1}{2}\left(\frac{8}{3}\right)$$

$$t = \frac{1}{2}\left(\frac{8}{3}\right)$$

$$t = \frac{4}{3}.$$

Exercise 2.3

Solve and check:

1. $10z = 40$

2. $25u = 400$

3. $15x = 9$

4. $18y = 27$

5. $2.5r = 10$

6. $0.3p = 0.27$

7. $\dfrac{1}{2}t = \dfrac{3}{4}$

8. $\dfrac{3}{8}s = \dfrac{3}{2}$

9. $\dfrac{y}{5} = 9$

10. $\dfrac{x}{33} = 3$

11. $\dfrac{r}{1.3} = 3$

12. $\dfrac{s}{0.4} = 11.2$

13. $\dfrac{z}{24} = \dfrac{5}{6}$

14. $\dfrac{w}{8} = \dfrac{3}{20}$

15. $\dfrac{5}{6}t = \dfrac{25}{9}$

16. $\dfrac{9}{10}u = \dfrac{36}{5}$

17. $\dfrac{3}{5}v = \dfrac{15}{2}$

18. $\dfrac{4}{5}y = \dfrac{15}{16}$

19. $\dfrac{3}{4}z = \dfrac{4}{3}$

20. $\dfrac{4}{9}w = \dfrac{4}{9}$

2.4   Applied Problems

Solving equations is useful in solving applied problems. An applied problem is a verbally stated problem that can be reduced to an equation. We solve the equation to solve the problem.

You may have seen percent problems in a basic math course. If you had not learned any algebra, you had to learn how to identify and solve several different cases of percent problems. Using just Method 3 of equation solving, all the different cases of basic percent problems can be reduced to one case, and the problem can be solved by solving an equation.

Basic percent problems have three parts. The **rate** is the percent. If we are taking 20% of 50, the rate is 20%. The **base** is the amount on which we take the percent. The base often can be identified as the amount following the word *of*. If we take 20% of 50, the base is 50. The **percentage** is the result; in our example, what we get when we take 20% of 50.

To find the percentage we must write the rate as a decimal. *Percent* means, literally, "per 100," or divided by 100. This is the same as moving the decimal point two places to the left. To write 20% as a decimal, we may write

$$20\% = \frac{20}{100} = 0.2.$$

To find 20% of 50, we write 20% as a decimal and multiply by 50:

$$(0.2)(50) = 10.$$

The percentage is 10.

EXAMPLE 2.18   Find 3.5% of 29.25.

Solution

First, we change the percent to a decimal by moving the decimal point two places to the left:

$$3.5\% = 0.035.$$

Then we multiply the rate times the base:

$$(0.035)(29.25) = 1.02375.$$

The percentage is 1.02375.

To find a percentage given the rate and base, we do not need algebra. However, if we are given the percentage and either the rate or the base, algebra can be a useful tool. In general, if R is the rate, B is the base, and P is the percentage,

$$RB = P.$$

Any of the three letters can be the variable. The other two will be replaced by constants, which are the given numerical values in the problem.

EXAMPLE 2.19   If 13.2% of a number is 20.3, what is the number?

Solution

In this example we are given the rate and the percentage. We must find the base. Writing 13.2% as a decimal, and using the equation

$$RB = P,$$

we have

$$0.132B = 20.3.$$

To solve the equation, we divide each side by 0.132:

$$\frac{0.132B}{0.132} = \frac{20.3}{0.132}$$

$$B = \frac{20.3}{0.132}$$

$$B = 153.787.$$

In this problem, each number had three digits. Rounding off to three digits,

$$B = 154.$$

To check, we find 13.2% of 154:

$$(0.132)(154) = 20.328$$

This result rounds off to 20.3.

EXAMPLE 2.20   What percent of 33.92 is 9.33?

Solution   In this example we are given the base and the percentage. Using

$$RB = P$$

we have

$$R(33.92) = 9.33,$$

or

$$33.92R = 9.33.$$

Dividing each side by 33.92,

$$\frac{33.92R}{33.92} = \frac{9.33}{33.92}$$

$$R = \frac{9.33}{33.92}$$

$$R = 0.275.$$

We have rounded R to three digits, the smaller of the numbers of digits in the given numbers. Observe that the rate R is written as a decimal. To write R as a percent, we move the decimal point back two places to the right:

$$R = 27.5\%.$$

You should check this solution by finding 27.5% of 33.92.

In applied problems, we must be careful to use the correct numbers for the base and the percentage. These numbers might not be the numbers given directly in the problem.

EXAMPLE 2.21   A watch originally cost $60. It is on sale for $48. What is the percent discount?

Solution   We are given the base, which is $60. However, we are not given the percentage directly. The percentage is the amount of the discount, not the sale price of $48. The amount of the discount is

$$60 - 48 = 12.$$

Therefore, the percentage is $12. We write

$$RB = P$$
$$R(60) = 12$$
$$\frac{60R}{60} = \frac{12}{60}$$
$$R = \frac{12}{60}$$
$$R = 0.2.$$

To write R as a percent, we move the decimal point two places to the right. The percent discount is 20%. To check, observe that, with a 20% discount, the amount you pay is 80% of the original price. Then check that 80% of $60 is $48.

EXAMPLE 2.22   A jacket has been marked down 30%. Its price is now $29.75. What was the original price of the jacket?

Solution   The price we are given, $29.75, is the percentage. However, we are not given the rate directly. Since $29.75 is the amount left when 30% is taken off, or 70% of the base, the rate is 70%. Then,

$$RB = P$$
$$0.70B = 29.75$$
$$\frac{0.70B}{0.70} = \frac{29.75}{0.70}$$
$$B = \frac{29.75}{0.70}$$
$$B = 42.5.$$

The original price of the jacket was $42.50. To check, you should calculate 70% of $42.50 to obtain $29.75. Alternatively, you may calculate 30% of $42.50 to obtain $12.75, which is $42.50 − $29.75, the amount of the discount.

Uniform motion problems involve a **distance** D, an average **rate** R, and a **time** T. Suppose you drive at an average rate of 30 miles per hour for 2 hours. We say an average rate because you will not be going 30 miles per hour at every instant. Sometimes you will go a little faster and sometimes a little slower. Sometimes you may be going 55 miles per hour on an expressway, and sometimes you may be stopped (0 miles per hour) at a traffic light. However, if you average 30 miles per hour for 2 hours, you will go

$$(30)(2) = 60,$$

a distance of 60 miles. In general,

$$RT = D.$$

EXAMPLE 2.23   A train travels 200 miles at an average rate of 40 miles per hour. How long does it take to make the trip?

Solution   We are given the distance and rate, and must find the time. Using

$$RT = D,$$

we have

$$40T = 200.$$

Dividing each side by 40,

$$\frac{40T}{40} = \frac{200}{40}$$
$$T = 5.$$

It will take 5 hours to make the trip. To check, you should calculate 40 miles per hour times 5 hours to obtain 200 miles.

EXAMPLE 2.24   A ball falls 49 feet in 1.75 seconds. What is the average rate at which it falls?

Solution   We are given the distance and time, and must find the rate. Using

$$RT = D,$$

we have

$$R(1.75) = 49$$

$$\frac{1.75R}{1.75} = \frac{49}{1.75}$$

$$R = 28.$$

The average rate is 28 feet per second. To check, you should calculate 28 feet per second times 1.75 seconds to obtain 49 feet. Observe, however, that the ball does not fall at 28 feet per second throughout the 1.75 seconds. It starts at a rate of 0 feet per second and gains speed as it falls.

Unit pricing is a convenient way to compare prices of items in a supermarket. Suppose an item is measured in ounces. The price of the item is the price per ounce times the number of ounces. For example, suppose a jar of peanut butter costs 6.05¢ an ounce and contains 28 ounces. The price of the jar of peanut butter is

$$(6.05)(28) = 169.4$$

The price is 169¢, or $1.69. In general, if U is the unit price, N is the number of units, and P is the price of the item,

$$UN = P.$$

EXAMPLE 2.25   A $14\frac{1}{2}$-ounce can of soup costs 57¢. What is the price per ounce?

Solution   We are given the price and number of ounces, and must find the unit price. Using

$$UN = P,$$

we have

$$U(14.5) = 57$$

$$\frac{14.5U}{14.5} = \frac{57}{14.5}$$

$$U = 3.93,$$

where we have rounded off to two decimal places. The unit price is 3.93¢ per ounce. To check, you should calculate 3.93¢ per ounce times 14.5 ounces to obtain 56.985, or 57¢.

Often, unit prices are given in pounds. There are 16 ounces in a pound.

EXAMPLE 2.26   A 10-ounce box of frozen vegetables costs 43¢. What is the unit price per pound?

Solution   First we find the price per ounce:

$$UN = P$$
$$U(10) = 43$$
$$\frac{10U}{10} = \frac{43}{10}$$
$$U = 4.3.$$

The unit price is 4.3¢ per ounce. Since there are 16 ounces in a pound, we calculate

$$(4.3)(16) = 68.8.$$

The unit price is 68.8¢ per pound. To check, we observe that an amount of 10 ounces is $\frac{10}{16}$ pound. Then, we calculate 68.8¢ per pound times $\frac{10}{16}$ pound:

$$(68.8)\left(\frac{10}{16}\right) = 43,$$

to obtain the result 43¢.

Exercise 2.4

1. Find 10.2% of 135.

2. Find 0.5% of 15.25.

3. 15% of a number is 18. What is the number?

4. 8.13 is 9.5% of a number. What is the number?

5. What percent of 148 is 24?

6. What percent of 133.3 is 114.3?

7. A bookcase that originally sold for $90.00 is on sale for $67.50. What is the percent discount?

8. A sweater is on sale for $13.30. The original price was $19.95. What is the percent discount?

9. A calculator is marked down 40%. What was its original price if the price is now $4.80?

10. After a discount of 16%, a piano costs $755.99. What was the original price?

11. A car travels 50 miles at an average rate of 30 miles per hour. How long does it take?

12. A plane flies 300 miles at an average rate of 240 miles per hour. How long does it take?

13. A boat sails 3 nautical miles in $\frac{1}{2}$ hour. What is its average rate?

14. A projectile travels 50 feet in 0.2 seconds. What is the average rate?

15. A 12-ounce can of corn costs 35¢. What is the unit price per ounce to two decimal places?

16. A 38-ounce jar of catsup costs $1.39. What is the unit price per ounce?

17. A 15-ounce package of cookies costs $1.09. What is the unit price per pound?

18. A $6\frac{1}{2}$-ounce can of tuna costs 99¢. What is the unit price per pound?

19. A 46-ounce can of juice costs 79¢. What is the unit price per quart? (There are 32 ounces in a quart.)

20. A 48-ounce jar of salad oil costs $1.99. What is the unit price per gallon? (There are 128 ounces in a gallon.)

Self-test

1. Determine whether each of these numbers is a solution of the equation $2x + 9 = 4x - 2$:

   a. $x = \dfrac{3}{4}$

   b. $x = \dfrac{11}{2}$

   1a. _____

   1b. _____

Solve and check:

2. $\dfrac{z}{5} = \dfrac{3}{10}$

   2. _____

3. $t - 1.8 = 0.3$

   3. _____

4. $u + \dfrac{2}{3} = \dfrac{4}{5}$

   4. _____

5. A radio is marked down 20% and now costs $34.00. What was the original price of the radio?

   5. _____

UNIT 3

The Integers

INTRODUCTION
In the preceding unit, you solved equations such as $x - 19 = 3$. However, suppose you must solve the equation $x + 19 = 3$. There is no positive number x that is the solution of this equation. In this unit you will learn about negative numbers, which provide solutions to such equations. You will learn about a set of negative and nonnegative numbers called the integers. The integers will be used to illustrate concepts such as the number line, order, and absolute value.

OBJECTIVES
When you have finished this unit you should be able to:
1. Graph a given integer on the number line and write the coordinate of a point indicated on the number line.
2. Given two integers, use the sumbols $<$ and $>$ to indicate which is the smaller and which the larger.
3. Find the absolute value of a given integer and use absolute value to find the distance from the origin of a point on the number line.

3.1   Definition of the Integers

The equation

$$x - 1 = 0$$

is very easy to solve. Adding 1 to both sides,

$$\begin{aligned} x - 1 &= 0 \\ +\, 1 &\quad +\, 1 \\ \hline x + 0 &= 1 \\ x &= 1. \end{aligned}$$

Clearly $x = 1$ is the solution since, substituting 1 for x,

$$\begin{aligned} x - 1 &= 0 \\ 1 - 1 &\overset{?}{=} 0 \\ 0 &= 0. \end{aligned}$$

However, with the numbers we have been using so far, we cannot solve such a seemingly simple equation as $x + 1 = 0$, or an equation such as $x + 19 = 3$. To solve such equations, we must be able to subtract a larger number from a smaller number.

It does not seem sensible to be able to solve an equation such as $x - 1 = 0$ but not $x + 1 = 0$, or an equation such as $x - 19 = 3$ but not $x + 19 = 3$. Therefore, numbers were invented to represent the solutions to equations such as $x + 1 = 0$ and $x + 19 = 3$. These numbers are called **negative** numbers. The numbers we have used so far, the **positive** numbers and **zero,** are called **nonnegative** numbers.

Positive numbers, including fractions and radicals, were used by ancient civilizations. There is evidence of fractions in both a Babylonian tablet and an Egyptian papyrus, each dating from about 1700 B.C. The Greek mathematician and inventor Archimedes studied square roots and found a good approximation of π as early as 200 B.C. Zero was recognized as a number, rather than simply a place-holder, about A.D. 500 by Hindu mathematicians, from whose symbols our numerals are descended. For centuries, however, negative numbers were not accepted, even by some of the greatest mathematicians.

The Hindus had interpreted negative numbers as representing debts by the seventh century. Negative numbers came to Europe by means of Arabian books, but negative solutions to equations were rejected. For instance, the great Italian equation solvers did not use negative solutions. It was not until early in the nineteenth century that the great German mathematician Karl Friedrich Gauss (1777–1855) showed the necessity of recognizing negative solutions to equations.

One who did accept negative numbers was the French philosopher and mathematician René Descartes (1596–1650). An interpretation of negative numbers has its roots in his work on coordinate geometry. Coordinates that are negative numbers were introduced soon after by an English mathematician, John Wallis (1616–1703).

The coordinate interpretation of negative numbers uses the **number line.** The number line associates every number—positive, zero, and negative—with a point on a line. The number is called the **coordinate** of the point.

To construct the number line, we first choose a **direction** for the line. The number line is usually drawn horizontally and directed to the right:

The arrow indicates the direction.

Next, we choose the point that will have the coordinate 0. This point is called the **origin:**

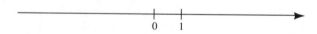

The origin may be anywhere we like on the line.

Finally, we choose the point that will have the coordinate 1. This point establishes the **unit:**

The unit may be any size we like. We must place 1 to the right of 0, however, because the direction of the line is to the right. Then, the point with coordinate 2 is one unit to the right of 1, the point with coordinate 3 is one unit to the right of 2, and so on:

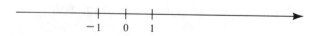

Now, we define our first negative number. The number **negative one** is the coordinate of the point one unit to the *left* of the origin. Negative one is written -1:

Negative two, negative three, and so on, are each placed one unit further to the left, and they have similar symbols:

Observe that, since 0 is to neither the right nor the left of the origin, we say that 0 is neither positive nor negative.

To match a point on the number line with its coordinate is to **graph** the point. When we graph a point, we indicate the point by a dot and write its coordinate below the dot.

The set of all numbers that have graphs on the number line is called the set of **real numbers.** Throughout this unit and the next, we will use a part of the set of real numbers called the **integers.** The set of integers may be written

$$\ldots, -5, -4, -3, -2, -1, 0, 1, 2, 3, 4, 5, \ldots$$

where three dots, called an _ellipsis,_ mean that the numbers continue indefinitely.

The numbers 1, 2, 3, 4, 5, . . . are the **natural numbers, or positive integers.** The numbers 0, 1, 2, 3, 4, 5, . . . are the whole numbers, or **nonnegative integers.** The numbers . . . , −5, −4, −3, −2, −1 are the negatives of the natural numbers, or **negative integers.**

EXAMPLE 3.1   Graph the point that has the coordinate:

a. 7   b. −3

Solutions   a. Since 7 is positive, we must go seven units to the right of the origin:

b. Since −3 is negative, we must go three units to the left of the origin:

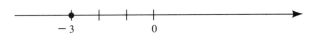

EXAMPLE 3.2   Write the coordinate of the point:

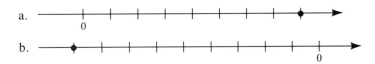

Solutions   a. Since the dot is eight units to the right of the origin, the coordinate is 8.
b. Since the dot is nine units to the left of the origin, the coordinate is −9.

Exercise 3.1

Graph the point that has the coordinate:

1. 1   2. −1   3. 2   4. −2

5. 0   6. 5   7. 12   8. −5

9. −8   10. −12

Write the coordinate of the point:

11.

12.

13.

14.

15.

16.

| 3.2 | # Order of the Integers |

Order relations compare the sizes of two numbers. You already know some order relations for the positive integers. If a and b represent positive integers, then a and b may be the same size. In this case we say

$$a = b.$$

For example, we could write

$$10 = 3 + 7$$

where a is 10 and b is $3 + 7$.

If a and b are not the same size, then one is smaller and one is larger. If a is smaller than b, we write

$$a < b \text{ (read ``a is less than b'').}$$

For example,

$$7 < 10.$$

If a is larger than b, we write

$$a > b \text{ (read ``a is greater than b'').}$$

For example,

$$7 > 3.$$

These concepts of order relations can be extended to all the integers. Mathematicians call the basic property of order relations the **trichotomy axiom.**

The Trichotomy Axiom: For any two integers a and b, exactly one of the following is true:

$$a = b$$
$$a < b$$
$$a > b$$

For any two integers a and b where $a \neq b$, we need a way to determine whether $a < b$ or $a > b$. Consider the integers as coordinates of points on the number line:

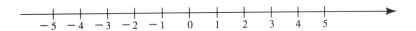

The direction of the number line was chosen to the right. This means that the coordinates are larger as we go to the right. For example, 5 is to the right of 4; therefore

$$5 > 4.$$

Similarly, 4 is to the left of 5; therefore,

$$4 < 5.$$

Observe that every positive integer is to the right of 0. If a is any positive integer, then

$$a > 0$$

or

$$0 < a.$$

Similarly, every negative integer is to the left of 0. If a is any negative integer, then

$$a < 0$$

or

$$0 > a.$$

EXAMPLE 3.3   Use $<$ or $>$ to indicate which is the larger integer and which is the smaller integer:

a. 4, 0   b. -7, 0

Solutions

a. Since any positive integer is greater than 0,

$$4 > 0.$$

b. Since any negative integer is less than 0,

$$-7 < 0.$$

Next, we observe that every positive integer is to the right of every negative integer. Therefore, if a is any positive integer and b is any negative integer,

$$a > b$$

or

$$b < a.$$

EXAMPLE 3.4   Use < or > to indicate which is the larger integer and which is the smaller integer:

a. 3, −5   b. −4, 6

Solutions

a. Since 3 is a positive integer and −5 is a negative integer,

$$3 > -5.$$

b. Since −4 is a negative integer and 6 is a positive integer,

$$-4 < 6.$$

Finally, we consider order relations comparing two negative integers. As with positive integers, the integer to the right is larger. For example, since −4 is to the right of −5,

$$-4 > -5.$$

Similarly, since −5 is to the left of −4,

$$-5 < -4.$$

It may seem strange that $-5 < -4$, whereas $5 > 4$. Think of it this way: If you have $5, you have more than $4; but, if you owe $5, you are further in debt than if you owed $4. Thus, −5 is more negative than −4.

EXAMPLE 3.5   Use < or > to indicate which is the larger integer and which is the smaller integer:

a. −10, −6   b. −7, −8

Solutions

a. Since −10 is to the left of −6 on the number line,

$$-10 < -6.$$

b. Since −7 is to the right of −8 on the number line,

$$-7 > -8.$$

Exercise 3.2

Use < or > to indicate which is the larger integer and which is the smaller integer:

1. 11, 5   2. 13, 15   3. 6, 0   4. 0, 5

5. −8, 0   6. 0, −4   7. 2, −7   8. −5, 9

9. $-4, 10$ 10. $4, -16$ 11. $-2, -7$ 12. $-4, -10$

13. $-8, -5$ 14. $-7, -6$ 15. $-15, -20$ 16. $-40, -36$

3.3   Absolute Value

There are times when it is important to know that one integer is to the left or right of another on the number line. The determination of order relations among the integers is such a situation. At other times, it does not matter whether an integer is to the left or right of another. In particular, we sometimes do not care if an integer is to the left or right of 0, but only how far it is from 0. In these situations we use the **absolute value** of the integer. The absolute value of any integer a is written

$$|a| \text{ (read ``the absolute value of a'')}.$$

Definition: If a is a positive integer, then

$$|a| = a.$$

If a is a positive integer, then $-a$ is a negative integer and

$$|-a| = a.$$

Also,

$$|0| = 0.$$

The absolute value of any integer is a nonnegative integer: the absolute value of any positive integer is a positive integer; the absolute value of any negative integer is a positive integer; and the absolute value of 0 is 0.

EXAMPLE 3.6   Find the value:

a. $|4|$ b. $|-4|$ c. $|0|$

Solutions

a. Since 4 is a positive integer,

$$|4| = 4.$$

b. Since -4 is a negative integer,

$$|-4| = 4.$$

c. For 0,

$$|0| = 0.$$

If absolute values of integers are combined with other integers, we must find the absolute values before combining the integers.

EXAMPLE 3.7   Find the value of $6|-10|$.

Solution   First, we find the absolute value:

$$6|-10| = 6(10)$$
$$= 60.$$

Observe that the absolute value symbol acted like multiplication parentheses. When we found the absolute value, we had to supply the multiplication parentheses.

The absolute value symbol may also act as grouping parentheses. In this case, we must combine the integers inside the absolute value symbol before finding the absolute value.

EXAMPLE 3.8   Find the value of $2|10 - 4|$.

Solution   First, we subtract 4 from 10:

$$2|10 - 4| = 2|6|$$
$$= 2(6)$$
$$= 12.$$

EXAMPLE 3.9   Find the value:

a. $|10 - 4|$ b. $|10| + |-4|$

Solutions   a. We must subtract 4 from 10 first:

$$|10 - 4| = |6|$$
$$= 6.$$

b. We must find the absolute values first and then combine the results:

$$|10| + |-4| = 10 + 4$$
$$= 14.$$

An important use of absolute value is in finding the *distance* of a point from the origin. If an integer a is the coordinate of a point on the number line, then the distance of the point from the origin is $|a|$.

EXAMPLE 3.10   Find the distance of the point from the origin:

Solutions

a. The coordinate of the point is 7. Therefore, the distance of the point from the origin is

$$|7| = 7.$$

b. The coordinate of the point is -8. Therefore, the distance of the point from the origin is

$$|-8| = 8.$$

It is a common error to confuse the coordinate of a point with its distance from the origin. The *coordinate* may be positive or negative, and whether it is positive or negative is determined by whether the point is to the right or to the left of the origin. The *distance* is never negative, regardless of whether the point is to the right or to the left of the origin. A point 8 units to the left of the origin is the same distance from the origin as a point 8 units to the right. Finally, observe that the coordinate of the origin itself is 0, and the distance of the origin from itself is $|0| = 0$.

Exercise 3.3

Find the value:

1. $|7|$

2. $|12|$

3. $|-7|$

4. $|-12|$

5. $|0|$

6. $3|0|$

7. $5|-8|$

8. $10|-10|$

9. $4|5 - 2|$

10. $5|16 - 7|$

11. $16|28 - 20|$

12. $12|200 - 100|$

13. $|8 - 5|$

14. $|8| + |-5|$

15. $|39| + |-9|$

16. $|39 - 9|$

17. $|22 - 22|$

18. $|22| + |-22|$

19. $|18| - |-18|$

20. $|-18| + |18|$

21. $|-32| + |20|$

22. $|-32| - |20|$

23. $|-14| - |-6|$

24. $|-14| + |-6|$

Find the distance of the point from the origin:

25.

26.

27.

28.

29.

30.

Self-test

1. Graph the point that has the coordinate:

   a. −6

   b. 4

2. Write the coordinate of the point:

   a.

   b.

   2a. _____

   2b. _____

3. Use < or > to indicate which is the larger integer and which is the smaller integer:

   a. −4, 0

   b. 5, −3

   c. −6, −4

   d. −5, −9

   3a. _____

   3b. _____

   3c. _____

   3d. _____

4. Find the value:

   a. $|9 - 2|$

   b. $|9| + |-2|$

   4a. _____

   4b. _____

5. Find the distance of the point from the origin:

   a.

   b.

   5a. _____

   5b. _____

Operations Involving Integers

INTRODUCTION

In the preceding unit you learned about the set of integers, consisting of the natural numbers, their negatives, and zero. You also learned some basic concepts related to the set of integers. In this unit you will learn how to do operations involving integers. First, you will learn how to do the four basic operations—addition, subtraction, multiplication, and division—for integers. Then, you will evaluate expressions involving two or more basic operations for integers.

OBJECTIVES

When you have finished this unit you should be able to:

1. Find the sum of two integers.
2. Find the product of two integers.
3. Find the quotient of two integers.
4. Find the difference of two integers.
5. Evaluate expressions involving sums, differences, products, and quotients of integers.

4.1 Addition

You are familiar with addition of natural numbers, which were defined as the positive integers in the preceding unit. Now, we consider the sum of two negative integers. Recall that we may think of the negative integers as representing a debt. Suppose you owe $5 and you also owe $3. Then you owe a total of $8. This is the same as saying

$$(-5) + (-3) = -8.$$

Observe that we have written the two negative integers in parentheses. The parentheses indicate that the minus symbols mean negative integers and not the minus sign in a subtraction. The operation is addition, indicated by the plus sign between the negative integers. We may also write

$$-5 + (-3) = -8$$

because the minus in front of the 5 could only indicate a negative integer.

We can use the number line to illustrate the addition of two negative integers. Again, consider the example $(-5) + (-3)$. To represent -5 on the number line, we go five units to the left of the origin. Then to add -3, we go three more units to the left:

We finish at the point corresponding to -8. Thus $(-5) + (-3) = -8$.

Now, observe the process of obtaining the preceding result. We have added the absolute values of the integers using the direction to the left to indicate the negative sign. We state the process as a rule for adding two negative integers.

Addition of Two Negative Integers: Add the absolute values and write the negative of the result.

EXAMPLE 4.1   Find the sum:

a. $(-6) + (-9)$ b. $-27 + (-13)$

Solutions   a. The absolute values are

$$|-6| = 6 \text{ and } |-9| = 9.$$

We add the absolute values to obtain

$$6 + 9 = 15.$$

Then, we write the negative of the result. Therefore,

$$(-6) + (-9) = -15.$$

b. The absolute values are

$$|-27| = 27 \text{ and } |-13| = 13.$$

We add the absolute values to obtain

$$27 + 13 = 40.$$

Then we write the negative of the result. Therefore,

$$(-27) + (-13) = -40.$$

If two integers have opposite signs, one positive and one negative, we may think of an asset and a debt. Suppose you have $8 and owe $3. Then you have $5 left. This is the same as saying

$$8 + (-3) = 5.$$

You might, instead, have $3 but owe $8. If you pay the $3, you will still owe $5. This is the same as saying

$$3 + (-8) = -5.$$

Again, we can use the number line to illustrate these additions. For $8 + (-3)$, we go eight units to the right and then three to the left, ending at 5:

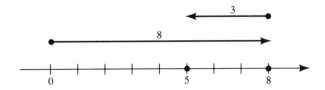

For $3 + (-8)$, we go three units to the right and then eight to the left, ending at -5:

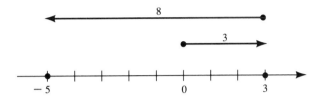

In each of these cases, we have subtracted the absolute values of the integers. In the first case, the absolute value of the positive integer is larger, and the result is positive. In the second case, the absolute value of the negative integer is larger and the result is negative. We state these processes as a rule for adding two integers with opposite signs.

Addition of Two Integers with Opposite Signs: Subtract the absolute values, and write the negative of the result if the absolute value of the negative integer is larger.

It is a common error to say that the negative integer is larger in the second case. But we cannot say, for example, that -8 is larger than 3 because any negative integer is less than any positive integer. However, we can say that $|-8| > |3|$.

EXAMPLE 4.2   Find the sum:

a. $16 + (-9)$ b. $8 + (-11)$

Solutions

a. The absolute values are

$$|16| = 16 \text{ and } |-9| = 9.$$

We subtract the absolute values to obtain

$$16 - 9 = 7.$$

Since $|16| > |-9|$, the result is positive. Therefore,

$$16 + (-9) = 7.$$

b. The absolute values are

$$|8| = 8 \text{ and } |-11| = 11.$$

We subtract the absolute values to obtain

$$11 - 8 = 3.$$

Then, since $|-11| > |8|$, we write the negative of the result. Therefore,

$$8 + (-11) = -3.$$

When the negative integer is written first, it is not necessary for either integer to be in parentheses, although one or both may be.

EXAMPLE 4.3   Find the sum:

a. $(-9) + 5$ b. $-24 + 43$

Solutions   a. The absolute values are

$$|-9| = 9 \text{ and } |5| = 5.$$

Subtracting the absolute values,

$$9 - 5 = 4.$$

Then since $|-9| > |5|$, we write the negative of the result. Therefore,

$$(-9) + 5 = -4.$$

b. The absolute values are

$$|-24| = 24 \text{ and } |43| = 43.$$

Subtracting the absolute values,

$$43 - 24 = 19.$$

Since $|43| > |-24|$, the result is positive. Therefore,

$$-24 + 43 = 19.$$

When zero is added to any integer, the result is the integer. For example, consider

$$(-14) + 0.$$

Although zero is neither positive nor negative, we will proceed as if the integers had opposite signs. The absolute values are

$$|-14| = 14 \text{ and } |0| = 0.$$

Subtracting the absolute values,

$$14 - 0 = 14.$$

Then since $|-14| > |0|$, we write the negative of the result:

$$(-14) + 0 = -14.$$

EXAMPLE 4.4   Find the sum of $0 + (-50)$.

Solution   When zero is added to any integer, the result is the integer. Therefore,

$$0 + (-50) = -50.$$

Finally, we consider two integers that have the same absolute value but opposite signs. The sum of two such integers is zero. For example, consider

$$7 + (-7).$$

The absolute values are the same:

$$|7| = 7 \text{ and } |-7| = 7.$$

Subtracting the absolute values,

$$7 - 7 = 0.$$

Therefore,

$$7 + (-7) = 0.$$

EXAMPLE 4.5   Find the sum of $(-50) + 50$.

Solution   The integers have the same absolute value but opposite signs. Therefore,

$$(-50) + 50 = 0.$$

Exercise 4.1

Find the sum:

1. $(-5) + (-7)$ 2. $(-11) + (-8)$ 3. $-25 + (-18)$ 4. $-48 + (-56)$

5. $6 + (-2)$ 6. $6 + (-10)$ 7. $(8) + (-13)$ 8. $(18) + (-9)$

9. $-14 + 7$ 10. $-1 + 9$ 11. $(-7) + 1$ 12. $(-19) + 30$

13. $(-8) + (15)$ 14. $(-15) + (12)$ 15. $-23 + 36$ 16. $-62 + 29$

17. $(-11) + 0$ 18. $0 + (-25)$ 19. $12 + (-12)$ 20. $-30 + 30$

4.2   Multiplication

Multiplication of integers can be defined in terms of addition. For example, (3)(6) means 6 taken 3 times:

$$(3)(6) = 6 + 6 + 6$$
$$= 18.$$

Similarly, $(3)(-6)$ means -6 taken 3 times:

$$(3)(-6) = (-6) + (-6) + (-6)$$
$$= -18.$$

On the number line, we go six units in the negative direction three times:

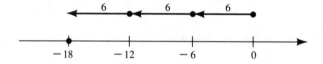

Also, we may interpret $(-3)(6)$ as -3 taken 6 times:

$$(-3)(6) = (-3) + (-3) + (-3) + (-3) + (-3) + (-3)$$
$$= -18.$$

On the number line, we go three units in the negative direction six times:

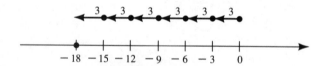

In each case, multiplying integers with opposite signs gives a negative integer.

> *Multiplication of Integers with Opposite Signs:*   Multiply the absolute values and write the negative of the result.

EXAMPLE 4.6   Find the product:

a. $(4)(-7)$ b. $(-8)(12)$

Solutions   a. We multiply the absolute values to obtain

$$(4)(7) = 28.$$

Then, we write the negative of the result. Therefore,

$$(4)(-7) = -28.$$

b. We multiply the absolute values to obtain

$$(8)(12) = 96.$$

Then, we write the negative of the result. Therefore,

$$(-8)(12) = -96.$$

It is not necessary to write the first factor of a multiplication in parentheses.

EXAMPLE 4.7 Find the product of $7(-3)$.

Solution Multiplying the absolute values and writing the negative of the result,

$$7(-3) = -21.$$

Observe that if the second factor is negative, as in this example, we must write the second factor in parentheses. If we write $7 - 3$, we appear to have the subtraction $7 - 3 = 4$ rather than a multiplication.

When an integer is multiplied by zero, the result is zero. For example, consider $(3)(0)$:

$$(3)(0) = 0 + 0 + 0$$
$$= 0.$$

EXAMPLE 4.8 Find the product of $0(-13)$.

Solution When an integer is multiplied by zero, the result is zero. Therefore,

$$0(-13) = 0.$$

Finally, we consider the product of two negative integers. For example, consider $(-3)(-6)$. To find this product, we begin with the fact that

$$-6 + 6 = 0.$$

We multiply each side of this equation by -3:

$$(-3)(-6 + 6) = (-3)(0).$$

We know that $(-3)(0) = 0$. Therefore,

$$(-3)(-6 + 6) = 0.$$

Now recall the distributive property from Section 1.4. We use the distributive property to write $(-3)(-6 + 6) = (-3)(-6) + (-3)(6)$. Thus,

$$(-3)(-6) + (-3)(6) = 0.$$

Finally, we know that $(-3)(6) = -18$, so

$$(-3)(-6) + (-18) = 0.$$

We also know that

$$18 + (-18) = 0.$$

Therefore,

$$
\begin{array}{r}
(-3)(-6) + (-18) = 0 \\
+\ 18 \quad +\ 18 \\
\hline
(-3)(-6) + 0 = 18 \\
(-3)(-6) = 18.
\end{array}
$$

To illustrate the preceding multiplication on the number line, we think of $(-3)(-6)$ as $-[3(-6)]$. We go six units in the negative direction three times. Then the negative in front reverses the direction. Thus, the result is to the right:

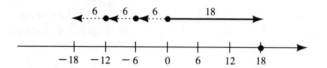

We see that the product of two negative integers is a positive integer.

Multiplication of Two Negative Integers: Multiply the absolute values.

EXAMPLE 4.9   Find the product:

a. $(-9)(-7)$ b. $-12(-21)$

Solutions

a. We multiply the absolute values to obtain

$$(9)(7) = 63.$$

Therefore,

$$(-9)(-7) = 63.$$

b. We multiply the absolute values to obtain

$$(12)(21) = 252.$$

Therefore,

$$-12(-21) = 252.$$

It is a common error to confuse addition of two negative integers and multiplication of two negative integers. The *sum* of two negative integers is *negative;* the *product* of two negative integers is *positive*.

EXAMPLE 4.10   Find the value of:

a. $(-16) + (-6)$ b. $(-16)(-6)$

Solutions   a. The operation is addition. We add the absolute values to obtain

$$16 + 6 = 22$$

and write the negative of the result:

$$(-16) + (-6) = -22.$$

b. The operation is multiplication, because there is no operation symbol between the sets of parentheses. We multiply the absolute values to obtain

$$(16)(6) = 96$$

and write

$$(-16)(-6) = 96.$$

Exercise 4.2

Find the product:

1. $(9)(-5)$

2. $(-6)(12)$

3. $8(-8)$

4. $-11(7)$

5. $(-4)(0)$

6. $(0)(-9)$

7. $0(-15)$

8. $-20(0)$

9. $(-5)(-13)$

10. $(-10)(-8)$

11. $-14(-6)$

12. $-15(-15)$

Find the value:

13. $(-12) + (-16)$

14. $-13 + (-7)$

15. $(-12)(-16)$

16. $-13(-7)$

17. $(-18) - (-8)$

18. $-9 - (-15)$

19. $(-18)(-8)$

20. $-9(-15)$

4.3   Division

In arithmetic, you learned that there is a relationship between division and multiplication. For example,

$$\frac{6}{3} = 2 \text{ because } 6 = (3)(2).$$

Now we consider $\frac{-6}{3}$. We know from Section 4.2 that $-6 = (3)(-2)$. Therefore,

$$\frac{-6}{3} = -2 \text{ because } -6 = (3)(-2).$$

Similarly, consider $\frac{6}{-3}$. We know from Section 4.2 that $6 = (-3)(-2)$. Therefore,

$$\frac{6}{-3} = -2 \text{ because } 6 = (-3)(-2).$$

Division of Integers with Opposite Signs: Divide the absolute values and write the negative of the result.

EXAMPLE 4.11   Find the quotient:

a. $\dfrac{-18}{6}$ b. $\dfrac{75}{-15}$

Solutions

a. We divide the absolute values to obtain

$$\frac{18}{6} = 3.$$

Then, we write the negative of the result. Therefore,

$$\frac{-18}{6} = -3.$$

b. We divide the absolute values to obtain

$$\frac{75}{15} = 5.$$

Then, we write the negative of the result. Therefore,

$$\frac{75}{-15} = -5.$$

To divide two negative integers, we may use a similar method. For example, consider $\frac{-6}{-3}$. We know from Section 4.2 that $-6 = (-3)(2)$. Therefore,

$$\frac{-6}{-3} = 2 \text{ because } -6 = (-3)(2).$$

The result is a positive integer.

Division of Two Negative Integers: Divide the absolute values.

EXAMPLE 4.12   Find the quotient of $\dfrac{-16}{-4}$.

Solution   We divide the absolute values to obtain

$$\frac{16}{4} = 4.$$

Therefore,

$$\frac{-16}{-4} = 4.$$

Divisions involving zero present some interesting problems. For example, consider $\frac{0}{6}$. We know from arithmetic that $0 = (6)(0)$. Therefore,

$$\frac{0}{6} = 0 \text{ because } 0 = (6)(0).$$

In general, for any integer a, where $a \neq 0$,

$$\frac{0}{a} = 0.$$

Now, consider $\frac{6}{0}$. We seek a number x such that

$$\frac{6}{0} = x \text{ because } 6 = (0)(x).$$

However, there is no such number x because, for any x, $0 = (0)(x)$.

Finally, consider $\frac{0}{0}$. We seek a number x such that

$$\frac{0}{0} = x \text{ because } 0 = (0)(x).$$

As we pointed out in the preceding case, $0 = (0)(x)$ is true for *any* number x. Thus division by zero is impossible because either there is no number, if the numerator is not zero, or there are infinitely many possible numbers, if the numerator and denominator both are zero. In either case, we say that division by zero is **undefined.**

Division Involving Zero: If a is an integer, $a \neq 0$, then

$$\frac{0}{a} = 0.$$

If a is an integer, including 0, then $\dfrac{a}{0}$ is undefined.

EXAMPLE 4.13   Find the quotient:

a. $\dfrac{0}{-4}$ b. $\dfrac{-4}{0}$

Solutions

a. Since $\dfrac{0}{a} = 0$ if $a \neq 0$,

$$\frac{0}{-4} = 0.$$

b. Since $\dfrac{a}{0}$ is undefined, $\dfrac{-4}{0}$ is undefined.

Exercise 4.3

Find the quotient:

1. $\dfrac{-9}{3}$ 2. $\dfrac{24}{-3}$ 3. $\dfrac{125}{-5}$ 4. $\dfrac{-50}{50}$

5. $\dfrac{-12}{-2}$ 6. $\dfrac{-52}{-26}$ 7. $\dfrac{-45}{-45}$ 8. $\dfrac{-1000}{-10}$

9. $\dfrac{0}{25}$ 10. $\dfrac{0}{-36}$ 11. $\dfrac{-25}{0}$ 12. $\dfrac{0}{0}$

4.4   Subtraction

Subtraction is related to addition in a way similar to the way division is related to multiplication. For example, we know from arithmetic that

$$5 - 3 = 2 \text{ because } 5 = 3 + 2.$$

We also know, from Section 4.1, that $5 + (-3) = 2$. Therefore,

$$5 - 3 = 5 + (-3).$$

Now, consider $5 - (-3)$. We know from Section 4.1 that $5 = (-3) + 8$. Therefore,

$$5 - (-3) = 8 \text{ because } 5 = (-3) + 8.$$

But we also know from arithmetic that $5 + 3 = 8$. Therefore,

$$5 - (-3) = 5 + 3.$$

Subtraction of an Integer: Add the integer with the opposite sign.

EXAMPLE 4.14   Find the difference:

a. $6 - 11$ b. $(-19) - 9$

Solutions   a. The integer following the subtraction symbol is positive. We add the corresponding negative integer:

$$6 - 11 = 6 + (-11)$$
$$= -5.$$

b. Again, the integer following the subtraction symbol is positive. We add the corresponding negative integer:

$$(-19) - 9 = (-19) + (-9)$$
$$= -28.$$

When a negative integer is written first, it need not be written in parentheses.

EXAMPLE 4.15   Find the difference of $-24 - 8$.

Solution   The integer following the subtraction symbol is positive. We add the corresponding negative integer:

$$-24 - 8 = -24 + (-8)$$
$$= -32.$$

EXAMPLE 4.16   Find the difference:

a. $7 - (-2)$ b. $(-4) - (-6)$

Solutions   a. The integer following the subtraction symbol is negative. We add the corresponding positive integer:

$$7 - (-2) = 7 + 2$$
$$= 9.$$

b. Again, the symbol following the subtraction symbol is negative. We add the corresponding positive integer:

$$(-4) - (-6) = (-4) + 6$$
$$= 2.$$

We may write the first integer in a subtraction with or without parentheses. However, if the second integer is negative, it must be written in parentheses.

EXAMPLE 4.17   Find the difference of $-18 - (-5)$.

Solution   Adding the corresponding positive integer,

$$-18 - (-5) = -18 + 5$$
$$= -13.$$

Observe that subtracting a negative integer is the same as adding the corresponding positive integer. Subtraction of a negative number is sometimes called a ''double negative.'' For example, $5 - (-3)$ involves subtracting a negative integer, which is a double negative. Since $5 - (-3) = 5 + 3$, you may have heard it said, ''a double negative is a positive.''

EXAMPLE 4.18   Find the value:

a. $-(-8)$ b. $-(-(-1))$

Solutions

a. We may think of $-(-8)$ as $0 - (-8)$. Then,

$$-(-8) = 0 - (-8)$$
$$= 0 + 8$$
$$= 8.$$

b. We find, as in part (a), that

$$-(-1) = 1.$$

Therefore, starting with the innermost parentheses,

$$-(-(-1)) = -(1)$$
$$= -1.$$

As with addition, we must be careful not to confuse subtraction with multiplication. We determine whether there is a minus sign indicating a subtraction or indicating a negative number.

EXAMPLE 4.19   Find the value:

a. $-16(-6)$ b. $-16 - (-6)$

Solutions

a. The operation is multiplication, since there is no operation symbol between the integer -16 and the parentheses enclosing the integer -6:

$$-16(-6) = 96.$$

b. There is a minus sign between the integer -16 and the parentheses enclosing the integer -6. Therefore, the operation is subtraction. We add the positive integer corresponding to -6:

$$-16 - (-6) = -16 + 6$$
$$= -10.$$

Subtraction of integers may be applied to finding distances between two points. In Section 3.3, we defined the distance of a point from the origin to be the absolute value of the coordinate of the point. Now, consider the distance between any two points on the number line. This distance is given by the absolute value of the difference of the coordinates.

EXAMPLE 4.20   Find the distance between the points on the number line with coordinnates 5 and -3.

Solution

We find the absolute value of the difference of the coordinates:

$$|5 - (-3)| = |5 + 3|$$
$$= |8|$$
$$= 8.$$

If we draw the number line and count the units between 5 and -3, we see that there are indeed 8 units between them:

The distance from -3 to 5 is of course the same:

$$|-3 - 5| = |-3 + (-5)|$$
$$= |-8|$$
$$= 8.$$

EXAMPLE 4.21   The high temperature one day is $11°$ above zero and the low is $2°$ below zero. What is the total change in temperature?

Solution   We may think of the thermometer as a vertical number line. We must find the distance between 11 and -2:

$$|11 - (-2)| = |11 + 2|$$
$$= |13|$$
$$= 13.$$

The total change is $13°$.

Exercise 4.4

Find the difference:

1. $19 - 7$

2. $6 - 29$

3. $(-11) - 3$

4. $(-4) - 14$

5. $-16 - 16$

6. $-9 - 1$

7. $9 - (-3)$

8. $15 - (-6)$

9. $(12) - (-1)$

10. $(22) - (-11)$

11. $(-5) - (-9)$

12. $(-16) - (-7)$

13. $-22 - (-14)$

14. $-24 - (-25)$

15. $0 - 12$

16. $-16 - 0$

Find the value:

17. $-(-6)$

18. $-(-20)$

19. $-(-(-12))$

20. $-(-(-(-16)))$

21. $-15(-4)$

22. $-2(-21)$

23. $-15 - (-4)$

24. $-2 - (-21)$

Find the distance between the points on the number line with coordinates:

25. 4 and -5

26. -2 and 8

27. 2 and 6

28. -3 and -9

Find the change in temperature between:

29. $10°$ above and $5°$ below zero.

30. $16°$ above and $7°$ below zero.

31. $23°$ and $3°$.

32. $1°$ below zero and $4°$ below zero.

Find the change in altitude between:

33. 100 feet above sea level and 60 feet below sea level.

34. 32 meters above sea level and 9 meters below sea level.

35. 14,000 feet above sea level and 22,000 feet above sea level.

36. 26 fathoms below sea level and 42 fathoms below sea level.

| 4.5 | # Expressions Involving the Integers

We will encounter expressions which involve more than one operation.

EXAMPLE 4.22   Find the value of $(-2) - (-7) + (-9)$.

Solution   Using subtraction of a negative integer,

$$(-2) - (-7) + (-9) = (-2) + 7 + (-9).$$

Then, combining the first two terms,

$$(-2) + 7 + (-9) = 5 + (-9).$$

Finally, combining the two remaining terms,

$$5 + (-9) = -4.$$

We will usually write such solutions with these steps:

$$(-2) - (-7) + (-9) = (-2) + 7 + (-9)$$
$$= 5 + (-9)$$
$$= -4.$$

Expressions may also contain grouping parentheses.

EXAMPLE 4.23   Find the value of $(-2) - 3(4 - 10)$.

Solution   Whenever possible, we simplify inside the parentheses first. Since $4 - 10 = -6$, we have

$$(-2) - 3(4 - 10) = (-2) - 3(-6).$$

We now have a subtraction and a multiplication. Recall from arithmetic the order of operations: it is agreed that we do multiplication before addition or subtraction:

$$(-2) - 3(-6) = (-2) - (-18)$$
$$= (-2) + 18.$$

Observe that, because the multiplication gave us a double negative, we now have an addition:

$$(-2) + 18 = 16.$$

We write this solution:

$$(-2) - 3(4 - 10) = (-2) - 3(-6)$$
$$= (-2) + 18$$
$$= 16.$$

If there are parentheses within parentheses, the outside parentheses may be written as brackets.

EXAMPLE 4.24   Find the value of $-5[(3 - 24) - (-6)]$.

Solution   We start with the innermost parentheses and work out to the brackets:

$$-5[(3 - 24) - (-6)] = -5[(-21) - (-6)]$$
$$= -5[(-21) + 6]$$
$$= -5[-15]$$
$$= 75.$$

The division line acts as a grouping symbol. We must do the operations above the line and the operations below the line before we can do the division.

EXAMPLE 4.25   Find the value of $\dfrac{-7 - 5}{8 - (-4)}$.

Solution   We do the operations above and below the division line first:

$$\frac{-7 - 5}{8 - (-4)} = \frac{-7 - 5}{8 + 4}$$
$$= \frac{-12}{12}$$
$$= -1.$$

Exercise 4.5

Find the value:

1. $3 - (6) - (-8)$

2. $-2 + (-7) - (-1)$

3. $-4 + (-10) - (-4)$

4. $-(-9) - (8) - (-2)$

5. $(-6) - 6(1 - 4)$

6. $-9 - 5(7 - 12)$

7. $6 - (-9 - 5)$

8. $3(6 - 12) - (-4)$

9. $-4[6 - (8 - 15)]$

10. $[(-24 + 16) - 10] - 9$

11. $8 - [0 - (-7 - 3)]$

12. $-14[-2 - (-11 + 9)]$

13. $\dfrac{4 - (-10)}{(-7) + 5}$

14. $\dfrac{0 - 16}{-2 - 6}$

15. $\dfrac{-3 - (-3)}{-14 - 18}$

16. $\dfrac{19 - (-19)}{-7 + 7}$

17. $(2 - 3)\left[\dfrac{6 - (10 - 2)}{12 - 10}\right]$

18. $[3 - (5 - 9)]\left[\dfrac{6 - 15}{12 - 15}\right]$

19. $[(-19) + (14 - 25)] - [(16 - 21) - 5]$

20. $[(17 - 31) - (42 - 21)] - [(33 - 21) - 12]$

Self-test

Find the value:

1. a. $-22 + (-17)$

    b. $-18 + 12$

    c. $-13 - 7$

    d. $8 - (-24)$

2. a. $(-6)(-7)$

    b. $\dfrac{40}{-5}$

    c. $(-9)(0)$

    d. $\dfrac{-12}{0}$

3. a. $-8(-11)$

    b. $-8 - (-11)$

4. $-2[-3 - (4 - 11)]$

5. The temperature drops from 24° to 7° below zero. What is the total change in temperature?

1a. _____

1b. _____

1c. _____

1d. _____

2a. _____

2b. _____

2c. _____

2d. _____

3a. _____

3b. _____

4. _____

5. _____

UNIT **5**

Cumulative Review

INTRODUCTION
In this unit you should make certain you remember all the material in all of the preceding units.

OBJECTIVE
When you have finished this unit you should be able to demonstrate that you can fulfill every objective of each preceding unit.

To prepare for this unit you should review the Self-Tests for Units 1 through 4. Do each problem of each Self-Test over again. If you cannot do a problem, or even if you have the slightest difficulty, you should:

1. Find out from the answer section the Objective for the unit to which the problem relates.
2. Review all the material in the section which has the same number as the objective, and redo all the Exercises for the section.
3. Try the Self-Test for the unit again.

Repeat these steps until you can do each problem in each Self-Test for each of Units 1 through 4 easily and accurately.

Self-test

1. Evaluate the expression

   $$xy + \frac{1}{2}z$$

   if $x = \frac{1}{6}$, $y = \frac{3}{4}$, and $z = \frac{5}{3}$.

   1. _____

2. Use the distributive property to remove the parentheses in the expression

   $$3\left(\frac{1}{2}s + \frac{1}{3}t\right) + 4\left(\frac{1}{8}u + \frac{1}{2}v\right)$$

   2. _____

Solve and check:

3. $y + \dfrac{2}{5} = \dfrac{7}{10}$

   3. _____

4. $16t = 56$

   4. _____

5. A $4\frac{1}{2}$-ounce can of shrimp costs \$2.09. What is the unit price per pound, to two decimal places?

   5. _____

6. Graph the point which has the coordinate:

   a. 5

   b. −2

Find the value:

7. $|-3| + |4|$

7. _____

8. a. $5 - 19$

   b. $5(-19)$

8a. _____

8b. _____

9. a. $\dfrac{-108}{-4}$

   b. $\dfrac{0}{-9}$

9a. _____

9b. _____

10. $-8[-3 - (4 + 0)]$

10. _____

UNIT

Simplifying Expressions

INTRODUCTION

In Unit 2, you learned four methods for solving equations, and you learned some applications of one form of equation. Most applications, however, involve equations containing more complicated expressions and also often involve negative numbers. To solve such equations, you must be able to simplify the expressions they contain, in order to use the methods for solving equations. In this unit you will learn how to simplify expressions such as those which might appear in equations.

OBJECTIVES

When you have finished this unit you should be able to:

1. Combine like terms of an expression.
2. Multiply an expression by a constant.
3. Find the sum or difference of two expressions.
4. Simplify a general linear expression.

6.1 Combining Like Terms

In Unit 1, and other units since, we have used the distributive property to remove parentheses. For example, using the distributive property, we may write

$$2(x + y) = 2x + 2y.$$

Now, suppose we begin with the expression

$$2x + 2y.$$

We may use the distributive property from right to left to write 2 as a **factor:**

$$2x + 2y = 2(x + y).$$

When we use the distributive property in this way, we **factor out a common factor.** In the preceding example, we have factored out the common factor 2.

A common factor may be a variable. Consider the expression

$$2x + 3x.$$

The common factor is x. If we factor out the common factor, we have

$$2x + 3x = x(2 + 3)$$
$$= x(5)$$
$$= 5x.$$

71

Observe that we have combined the two terms of the given expression. Terms such as $2x$ and $3x$, which can be combined, are called **like terms.**

If we have an expression such as

$$2x + 3y,$$

we find no common factor. Therefore, we cannot combine the two terms. The terms $2x$ and $3y$ are not like terms.

Definition: Two or more terms are like terms if all of their literal parts are identical.

EXAMPLE 6.1   Determine whether the terms are like terms:

a. $2ax$ and $-5ax$.   b. $3ay$ and $4by$.   c. x^2 and $2x$.

Solutions

a. The literal parts, ax, are the same in each term. Therefore, the terms are like terms.
b. Some literal parts, a and b, are not identical. Therefore, the terms are not like terms.
c. Although the variable x is the same in each term, the literal parts x^2 and x are not identical. Therefore, the terms are not like terms.

Whenever we have like terms, we can combine the terms. If we do not have like terms, we cannot combine them.

EXAMPLE 6.2   Combine like terms of $5u + 3u$.

Solution

Using the distributive property to factor out the common factor, we may write

$$5u + 3u = u(5 + 3)$$
$$= u(8)$$
$$= 8u.$$

Usually, we will skip the middle steps and add the numerical coefficients directly:

$$5u + 3u = 8u.$$

EXAMPLE 6.3   Combine like terms of $11s + 5s$.

Solution

We observe that the terms are like terms, so we may add the numerical coefficients:

$$11s + 5s = 16s.$$

EXAMPLE 6.4   Combine like terms of $9s + 6t$.

Solution

$9s$ and $6t$ are not like terms and so they cannot be combined. We must leave the expression in the form $9s + 6t$.

We may combine like terms that are subtracted by methods similar to those for addition. Recall from Section 4.4 that subtraction is the same as addition with the opposite sign. We use this fact to derive the distributive property for multiplication over subtraction:

$$x(y - z) = x[y + (-z)]$$
$$= xy + x(-z)$$
$$= xy - xz.$$

EXAMPLE 6.5   Combine like terms:

a. $7x - 4x$ b. $z - 6z$

Solutions

a. We use the distributive property for multiplication over subtraction to factor out the common factor x:

$$7x - 4x = x(7 - 4)$$
$$= x(3)$$
$$= 3x.$$

Usually, we will skip the middle steps and subtract the numerical coefficients directly:

$$7x - 4x = 3x.$$

b. We recall that $z = 1z$:

$$z - 6z = 1z - 6z$$
$$= z(1 - 6)$$
$$= z(-5)$$
$$= -5z,$$

or, subtracting the numerical coefficients directly,

$$z - 6z = -5z.$$

An expression may have more than two terms. Some terms within the expression may be like terms and others not like terms.

EXAMPLE 6.6   Combine like terms of $9y - 4y - 8y$.

Solution

All the terms are like terms. We may write

$$9y - 4y - 8y = 5y - 8y$$
$$= -3y.$$

EXAMPLE 6.7   Combine like terms of $ax - 5bx + 10ax$.

Solution

The terms ax and $10ax$ are like terms. Therefore, we write

$$ax - 5bx + 10ax = 11ax - 5bx.$$

However, $11ax$ and $-5bx$ are not like terms. Thus, we must leave the expression in the form $11ax - 5bx$.

EXAMPLE 6.8   Combine like terms of $x - 2xy - 3yx + 6y$.

Solution   Observe that xy is the same as yx, so $-2xy$ and $-3yx$ are like terms. We may write $-3yx$ as $-3xy$:

$$x - 2xy - 3yx + 6y = x - 2xy - 3xy + 6y$$
$$= x - 5xy + 6y.$$

Exercise 6.1

Combine like terms:

1. $5y + 6y$

2. $12p + 7p$

3. $9z + z$

4. $15u + u$

5. $3x + 4y$

6. $8y - 2z$

7. $12p - 6p$

8. $9r - 10r$

9. $-4x - 5x$

10. $11s - s$

11. $3v + 4v - 6v$

12. $6t - 13t + 3t$

13. $2x - 3y + 4x$

14. $4x - 6x - 10$

15. $2ax - 3bx - 4bx + 12$

16. $2by + 2y - 3y - 3by$

17. $x - 2xy - yx + 2y$

18. $6x - 4xy + 9yx - 6y$

19. $ab + ba - 3ab - 4ba$

20. $3abc - 4bca + 5cab$

6.2   Multiplication by a Constant

To multiply an expression by a constant, we use the distributive property.

EXAMPLE 6.9   Multiply $6(5x + 2)$.

Solution   Using the distributive property,

$$6(5x + 2) = 6(5x) + 6(2)$$
$$= 30x + 12.$$

If a term is subtracted, we may multiply similarly, using the distributive property for multiplication over subtraction.

EXAMPLE 6.10   Multiply $3(4y - 10)$.

Solution   Using the distributive property for multiplication over subtraction,

$$3(4y - 10) = 3(4y) - 3(10)$$
$$= 12y - 30.$$

We may use the extended distributive property when an expression has more than two terms.

EXAMPLE 6.11 Multiply $7(2x - 3y - 4)$.

Solution Using the extended distributive property for multiplication over subtraction,

$$7(2x - 3y - 4) = 7(2x) - 7(3y) - 7(4)$$
$$= 14x - 21y - 28.$$

EXAMPLE 6.12 Multiply $-4(2s - 5)$.

Solution We use the distributive property for multiplication over subtraction, and the rules for multiplication of negative integers:

$$-4(2s - 5) = -4(2s) - (-4)(5)$$
$$= -8s - (-20)$$
$$= -8s + 20.$$

Observe that the sign preceding each term of the expression changes. When an expression is multiplied by a negative constant, each positive coefficient becomes negative and each negative coefficient becomes positive.

Removing parentheses from an expression preceded by a negative is the same as multiplying the expression by -1.

EXAMPLE 6.13 Remove the parentheses from $-(3t - 10)$.

Solution We multiply by -1:

$$-(3t - 10) = -1(3t - 10)$$
$$= -1(3t) - (-1)(10)$$
$$= -3t - (-10)$$
$$= -3t + 10.$$

Observe that the result is the same as if we had simply changed the sign preceding each term. The negative of an expression is the same expression with the opposite sign preceding each term. Usually, we will skip the middle steps and change the sign of each term directly:

$$-(3t - 10) = -3t + 10.$$

EXAMPLE 6.14 Remove the parentheses from $-(-3x - y + 14)$.

Solution Changing the sign of each term,

$$-(-3x - y + 14) = 3x + y - 14.$$

Exercise 6.2

Multiply:

1. $4(3x + 5)$

2. $9(2y + 6)$

3. $7(4z - 2)$

4. $8(3 - t)$

5. $3(9p - 10q)$

6. $12(r - 4s)$

7. $5(2x - 3y - 10)$

8. $11(4u + 6v - 11)$

9. $-6(4s - 5)$ 10. $-5(12t - 5)$ 11. $-12(12 - 3t)$ 12. $-9(8x - 7y)$

13. $-8(3x - 4z - 6)$ 14. $-10(x - 2y + 3w)$ 15. $-1(4r - 9)$ 16. $-1(12 - 16y)$

Remove the parentheses:

17. $-(6x - 3)$ 18. $-(-3 - 4y)$

19. $-(2x + 4y - 9)$ 20. $-(-5x - 6y + 10)$

6.3   Addition and Subtraction

Consider the expression

$$(2x + 3) + (4x - 5).$$

This expression represents the sum of the two expressions $2x + 3$ and $4x - 5$. To add the expressions, we remove the parentheses and combine like terms.

EXAMPLE 6.15   Add $(2x + 3) + (4x - 5)$.

Solution   First, we remove the parentheses:

$$(2x + 3) + (4x - 5) = 2x + 3 + 4x - 5.$$

Then, we combine the like terms $2x$ and $4x$, and also the like terms 3 and -5:

$$2x + 3 + 4x - 5 = 2x + 4x + 3 - 5$$
$$= 6x - 2.$$

EXAMPLE 6.16   Add $(5x - 6) + (4y - 7)$.

Solution   Removing the parentheses,

$$(5x - 6) + (4y - 7) = 5x - 6 + 4y - 7$$
$$= 5x + 4y - 6 - 7.$$

$5x$ and $4y$ are not like terms and cannot be combined. Combining -6 and -7,

$$5x + 4y - 6 - 7 = 5x + 4y - 13.$$

If one expression is to be subtracted from another, we must be careful when removing the parentheses. In Section 6.2, we saw that the negative of an expression is the same expression with the sign of each term changed. Therefore, to subtract an expression, we change the sign of each term and add the resulting expression.

EXAMPLE 6.17   Subtract $(2x + 3) - (4x - 5)$.

Solution   We are subtracting the expression $4x - 5$. Therefore, we change the sign of each term of $4x - 5$, and add $-4x + 5$:

$$(2x + 3) - (4x - 5) = 2x + 3 - 4x + 5$$
$$= 2x - 4x + 3 + 5$$
$$= -2x + 8.$$

EXAMPLE 6.18   Subtract $(5x - 6) - (4y - 7)$.

Solution   Changing the sign of each term of the expression we are subtracting,

$$(5x - 6) - (4y - 7) = 5x - 6 - 4y + 7$$
$$= 5x - 4y - 6 + 7$$
$$= 5x - 4y + 1.$$

EXAMPLE 6.19   Subtract $9 - (-3t + 2)$.

Solution   Changing the sign of each term of the expression we are subtracting,

$$9 - (-3t + 2) = 9 + 3t - 2$$
$$= 9 - 2 + 3t$$
$$= 7 + 3t$$

or, writing the term containing the variable first,

$$7 + 3t = 3t + 7.$$

EXAMPLE 6.20   Subtract $(-9r - 8) - (9 - 8r)$.

Solution   Changing the sign of each term of the expression we are subtracting,

$$(-9r - 8) - (9 - 8r) = -9r - 8 - 9 + 8r$$
$$= -9r + 8r - 8 - 9$$
$$= -r - 17.$$

Exercise 6.3

Add or subtract as indicated:

1. $(7x + 3) + (5x - 8)$

2. $(4t - 9) + (5t - 9)$

3. $(6r - 2) + (2 - 8r)$

4. $(14 - 5z) + (6 + 5z)$

5. $(3y - 5) + (2z - 6)$

6. $(9x - 7y) + (6y - 4z)$

7. $(10x - 4) - (6x - 9)$

8. $(3s - 13) - (4s + 8)$

9. $(y - 6) - (6y - 6)$

10. $(11v + 3) - (11v - 3)$

11. $(2u - 3) - (8u + 12)$

12. $(8 - 5s) - (7 - 4t)$

13. $3x - (5 - 2x)$

14. $16 - (-4y - 7)$

15. $(-5x - 11) - (-9x - 8)$

16. $(12 - 5r) - (-4r - 12)$

17. $(x - y + 3) + (x - 3y - 4)$

18. $xy + 3x - 2) - (xy - 3x + 1)$

19. $(2x - 2xy + 2y) - (2x - xy - 2y)$

20. $(3x - xy + 2y) - (2x - 3xy + y)$

6.4   General Linear Expressions

To reduce expressions in an equation to a form that we can use to solve the equation, we may use one or all of the methods in the preceding sections. First, we combine any like terms that may appear within parentheses. Then, we remove all parentheses by multiplying by a constant, adding, or subtracting. Finally, we combine any like terms which result. We refer to all of these techniques as the process of **simplifying an expression.**

EXAMPLE 6.21   Simplify $-6(-3t + 4t - 1)$.

Solution   First, we combine like terms within the parentheses:

$$-6(-3t + 4t - 1) = -6(t - 1).$$

Then, we multiply by the constant:

$$-6(t - 1) = -6t + 6.$$

EXAMPLE 6.22   Simplify $5(2x - 3) - 2(3x - 4)$.

Solution   There are no like terms within any set of parentheses. Therefore, we proceed to multiply by the constants to remove the parentheses:

$$5(2x - 3) - 2(3x - 4) = 10x - 15 - 6x + 8.$$

Then we combine the resulting like terms:

$$10x - 15 - 6x + 8 = 4x - 7.$$

It is important to distinguish carefully between multiplication by a constant and subtraction.

EXAMPLE 6.23   Simplify $-3(3z - 10)$.

Solution   Since the constant -3 is next to the parentheses, this example is a multiplication by a constant:

$$-3(3z - 10) = -9z + 30.$$

EXAMPLE 6.24   Simplify $-3 - (3z - 10)$.

Solution   There is a minus sign indicating subtraction between the constant -3 and the parentheses. Therefore, this example is a subtraction. We change the sign of each term:

$$-3 - (3z - 10) = -3 - 3z + 10$$
$$= -3z + 7.$$

EXAMPLE 6.25   Simplify $(s + 2) + 2(2s - 3)$.

Solution   Removing the parentheses and combining like terms,

$$
\begin{aligned}
(s + 2) + 2(2s - 3) &= s + 2 + 4s - 6 \\
&= 5s - 4.
\end{aligned}
$$

EXAMPLE 6.26   Simplify $4(6r - 8) - (5 - 12r)$.

Solution   Removing the parentheses and combining like terms,

$$
\begin{aligned}
4(6r - 8) - (5 - 12r) &= 24r - 32 - 5 + 12r \\
&= 36r - 37.
\end{aligned}
$$

Exercise 6.4

Simplify:

1. $-2(6y + 4y - 12y)$

2. $-5(3z - 8 - 8z)$

3. $3(4s - 6) - 4(5s - 8)$

4. $-6(3x - 2) - 2(4x + 12)$

5. $-4(-3t + 2)$

6. $8(y - 8)$

7. $-4 - (-3t + 2)$

8. $8y - (y - 8)$

9. $-9(x - 3) + 3$

10. $-7(-3z - 5) - 10$

11. $(r - 4) - 3(3r - 12)$

12. $11(2s + 6) + (3s + 4)$

13. $-5(4p - 3) - (8p - 15)$

14. $4(2u - 6) - (8u + 6)$

15. $-(6y + 5) - 5(6y + 6)$

16. $-(8 - 5x) - 2(10x - 4)$

Self-test

Simplify:

1. $3u - 15u + 8u$

2. $5x - 6xy + 8yx - 2y$

3. $9t - (3 - 4t)$

4. $-5(3x - y + 4)$

5. $6(3x - 2) - 5(5x + 1)$

1. _____

2. _____

3. _____

4. _____

5. _____

UNIT

7 Linear Equations

INTRODUCTION

Many applications of algebra involve linear equations in one variable. In this unit you will learn how to solve some types of linear equations in one variable. You will use the techniques for simplifying expressions and the four basic methods for solving equations.

OBJECTIVES

When you have finished this unit you should be able to:

1. Solve linear equations of the forms $ax + b = c$ and $ax + b = cx + d$.
2. Solve linear equations by simplifying one or more of the linear expressions.
3. Solve linear equations where one or more expressions have denominators that are constants.

7.1 Solving Linear Equations

In Unit 2, we used four methods for deriving equivalent equations:

1. Subtract the same number from both sides of the equation.
2. Add the same number to both sides of the equation.
3. Divide both sides of the equation by the same number (where the number is not zero).
4. Multiply both sides of the equation by the same number (where the number is not zero).

The equations we solved in Unit 2 each could be solved with the use of just one of these rules. In this section we will solve equations where two or more rules must be used. Also, in Unit 2 we used only positive numbers. Now, of course, our constants and solutions may include positive or negative numbers.

Consider an equation of the form $ax + b = c$. The variable is x, and a, b, and c will be replaced by constants, which may be positive or negative. To solve an equation of this form we must use subtraction or addition, and also division or multiplication. Usually, it is most efficient to use subtraction or addition first.

EXAMPLE 7.1 Solve $2x + 3 = 15$.

Solution First, we isolate the term $2x$ by subtracting 3 from both sides:

$$
\begin{array}{rcr}
2x + 3 &=& 15 \\
- 3 && - 3 \\
\hline
2x + 0 &=& 12 \\
2x &=& 12.
\end{array}
$$

Then, we isolate x by dividing by 2:

$$2x = 12$$
$$\frac{2x}{2} = \frac{12}{2}$$
$$1x = 6$$
$$x = 6.$$

The solution is 6. To check, we substitute 6 for x in the original equation:

$$2x + 3 = 15$$
$$2(6) + 3 \overset{?}{=} 15$$
$$12 + 3 \overset{?}{=} 15$$
$$15 = 15.$$

EXAMPLE 7.2   Solve $4x - 10 = 6$.

Solution

First, we isolate the term $4x$ by adding 10 to both sides:

$$
\begin{array}{rr}
4x - 10 = & 6 \\
+\ 10 & +\ 10 \\
\hline
4x +\ \ 0 = & 16 \\
4x = & 16.
\end{array}
$$

Then we isolate x by dividing by 4:

$$4x = 16$$
$$\frac{4x}{4} = \frac{16}{4}$$
$$1x = 4$$
$$x = 4.$$

The solution is 4. You should check this solution by substituting 4 for x in the original equation.

Whenever we subtract or add the appropriate number to isolate a term, we will obtain that term plus zero. Therefore, we will usually skip the third line shown in each of the two preceding examples. Similarly, whenever we divide or multiply by the appropriate number to isolate a variable, we will obtain one times that variable. Therefore, we will usually skip the next to last line shown in the two preceding examples.

EXAMPLE 7.3   Solve $5z + 21 = 1$.

Solution

To isolate the term $5z$, we subtract 21 from both sides:

$$
\begin{array}{rr}
5z + 21 = & 1 \\
-\ 21 & -\ 21 \\
\hline
5z \quad = & -20.
\end{array}
$$

To isolate z, we divide both sides by 5:

$$5z = -20$$

$$\frac{5z}{5} = \frac{-20}{5}$$
$$z = -4.$$

The solution is -4. To check, we substitute -4 for z:

$$5z + 21 = 1$$
$$5(-4) + 21 \stackrel{?}{=} 1$$
$$-20 + 21 \stackrel{?}{=} 1$$
$$1 = 1.$$

EXAMPLE 7.4 Solve $\dfrac{t}{4} - 8 = -5$.

Solution To isolate the term $\dfrac{t}{4}$, we add 8 to both sides:

$$\frac{t}{4} - 8 = -5$$
$$\underline{+ 8 \quad + 8}$$
$$\frac{t}{4} \quad = \quad 3.$$

To isolate t, we multiply both sides by 4:

$$\frac{t}{4} = 3$$
$$4\left(\frac{t}{4}\right) = 4(3)$$
$$t = 12.$$

The solution is 12. You should substitute 12 for t to check this solution.

A similar type of equation has the form $ax + b = cx + d$. Here, we must subtract or add to collect the terms involving variables as well as the constants. We do all subtractions and additions first. Then, we divide or multiply to isolate the variable.

EXAMPLE 7.5 Solve $6x + 11 = 2x - 13$.

Solution First, we collect the terms involving x by subtracting $2x$ from both sides:

$$6x + 11 = 2x - 13$$
$$\underline{-2x \qquad\quad -2x}$$
$$4x + 11 = \qquad -13.$$

Next, we collect the constants by subtracting 11 from both sides:

$$4x + 11 = -13$$
$$\underline{\quad -11 \quad -11}$$
$$4x \quad = -24.$$

Observe that, if we collect the terms involving variables on the left-hand side, we should collect the constants on the right-hand side.

Finally, we isolate x by dividing both sides by 4:

$$4x = -24$$
$$\frac{4x}{4} = \frac{-24}{4}$$
$$x = -6.$$

The solution is -6. To check this solution, we substitute -6 for x wherever x appears in the original equation:

$$6x + 11 = 2x - 13$$
$$6(-6) + 11 \overset{?}{=} 2(-6) - 13$$
$$-36 + 11 \overset{?}{=} -12 - 13$$
$$-25 = -25.$$

It is sometimes convenient to collect the terms involving variables on the right-hand side of the equation, and the constants on the left-hand side.

EXAMPLE 7.6   Solve $4y + 3 = 5y - 2$.

Solution   We may avoid dealing with negative numbers by subtracting $4y$ from both sides of the equation:

$$
\begin{array}{rcr}
4y + 3 = & & 5y - 2 \\
\underline{-\,4y} & & \underline{-\,4y} \\
3 = & & y - 2.
\end{array}
$$

Then, we add 2 to each side:

$$
\begin{array}{rcr}
3 = & y - 2 \\
\underline{+\,2} & \underline{+\,2} \\
5 = & y.
\end{array}
$$

Of course, $5 = y$ means the same as $y = 5$, so the solution is 5. You should check this solution in the original equation.

We could also solve the equation by collecting the terms involving y on the left-hand side but this method is slightly longer:

$$
\begin{array}{rcr}
4y + 3 = & & 5y - 2 \\
\underline{-\,5y} & & \underline{-\,5y} \\
-y + 3 = & & -2.
\end{array}
$$

Now, we must collect the constant terms on the right-hand side:

$$
\begin{array}{rcr}
-y + 3 = & -2 \\
\underline{-\,3} & \underline{-3} \\
-y = & -5.
\end{array}
$$

To complete the solution, we multiply both sides by -1:

$$-y = -5$$
$$-1(-y) = -1(-5)$$
$$y = 5.$$

EXAMPLE 7.7   Solve $3 - 2x = x + 7$.

Solution   We may add $2x$ to both sides to collect the terms involving x on the right-hand side:

$$\begin{aligned} 3 - 2x &= \quad x + 7 \\ + 2x \quad &\quad + 2x \\ \hline 3 \quad\; &= \quad 3x + 7. \end{aligned}$$

Now, we collect the constants on the left-hand side:

$$\begin{aligned} 3 &= 3x + 7 \\ -7 &\quad\quad -7 \\ \hline -4 &= 3x. \end{aligned}$$

Finally, dividing both sides by 3:

$$-4 = 3x$$
$$\frac{-4}{3} = \frac{3x}{3}$$
$$-\frac{4}{3} = x$$

or

$$x = -\frac{4}{3}.$$

The solution is $-\frac{4}{3}$. To check, we substitute $-\frac{4}{3}$ for x in the original equation and calculate each side separately using a common denominator:

$$3 - 2x = x + 7$$
$$3 - 2\left(-\frac{4}{3}\right) \stackrel{?}{=} -\frac{4}{3} + 7$$
$$3 + \frac{8}{3} \stackrel{?}{=} -\frac{4}{3} + 7$$
$$\frac{9}{3} + \frac{8}{3} \stackrel{?}{=} -\frac{4}{3} + \frac{21}{3}$$
$$\frac{9 + 8}{3} \stackrel{?}{=} \frac{-4 + 21}{3}$$
$$\frac{17}{3} = \frac{17}{3}.$$

Exercise 7.1

Solve and check:

1. $x - 3 = 5$ 2. $y + 10 = 4$ 3. $z + 1 = 3$ 4. $t - 12 = -15$

5. $\frac{s}{4} = 2$ 6. $3r = 15$ 7. $-5p = 20$ 8. $-\frac{q}{3} = 3$

9. $4x + 9 = 13$ 10. $5s - 6 = 14$ 11. $7t + 23 = 2$ 12. $4u + 45 = 21$

13. $3x + 2 = 10$ 14. $10z - 3 = 6$ 15. $\dfrac{y}{5} - 9 = -3$ 16. $\dfrac{z}{2} + 15 = 10$

17. $9y - 5 = 3y + 13$ 18. $10s - 4 = 8s + 3$ 19. $3r + 9 = 4r + 10$ 20. $6z - 15 = 10z - 9$

21. $5 - 3t = t + 8$ 22. $6 - x = x + 7$ 23. $5 - 4y = -1$ 24. $9 - 8x = -9$

7.2   General Linear Equations

A general linear equation consists of expressions that are general linear expressions. Recall from Unit 6 that simplifying a general linear expression may involve combining like terms, multiplication by a constant, addition or subtraction of linear expressions, or any combination of these operations. To solve a general linear equation, we first use these techniques to simplify any general linear expressions that appear in the equation.

EXAMPLE 7.8   Solve $3z + 6 - 4z = z + 4$.

Solution   First, we combine the like terms on the left-hand side:

$$3z + 6 - 4z = z + 4$$
$$6 - z = z + 4.$$

Now, we have an equation of the type in Section 7.1. To solve this equation, we will first add z to both sides and then subtract 4 from both sides:

$$
\begin{array}{rcl}
6 - z & = & z + 4 \\
\underline{+\ z} & & \underline{+\ \ z} \\
6 & = & 2z + 4 \\
\underline{-4} & & \underline{-\ 4} \\
2 & = & 2z \\
\dfrac{2}{2} & = & \dfrac{2z}{2} \\
1 & = & z
\end{array}
$$

or

$$z = 1.$$

The solution is 1. To check, we substitute 1 for z wherever z appears in the original equation:

$$3z + 6 - 4z = z + 4$$
$$3(1) + 6 - 4(1) \overset{?}{=} 1 + 4$$
$$3 + 6 - 4 \overset{?}{=} 5$$
$$9 - 4 \overset{?}{=} 5$$
$$5 = 5.$$

Writing additions or subtractions in the format we have used so far often becomes unwieldy. For the rest of this unit, we will write additions or subtractions on one line. For example, adding z to both sides,

$$6 - z = z + 4$$
$$6 - z + z = z + 4 + z$$
$$6 = 2z + 4,$$

and subtracting 4 from both sides,

$$6 = 2z + 4$$
$$6 - 4 = 2z + 4 - 4$$
$$2 = 2z.$$

Observe that we have shown the additions or subtractions on a single line in the middle step.

EXAMPLE 7.9   Solve $2u - 3 + u = 4 - u + 2$.

Solution   First, we combine the like terms on each side:

$$2u - 3 + u = 4 - u + 2$$
$$3u - 3 = 6 - u.$$

We add u to both sides:

$$3u - 3 = 6 - u$$
$$3u - 3 + u = 6 - u + u$$
$$4u - 3 = 6.$$

Then, we add 3 to both sides:

$$4u - 3 = 6$$
$$4u - 3 + 3 = 6 + 3$$
$$4u = 9$$
$$\frac{4u}{4} = \frac{9}{4}$$
$$u = \frac{9}{4}.$$

The solution is $\frac{9}{4}$. To check, we substitute $\frac{9}{4}$ for u wherever u appears in the original equation and use a common denominator separately on each side:

$$2u - 3 + u = 4 - u + 2$$
$$2\left(\frac{9}{4}\right) - 3 + \frac{9}{4} \stackrel{?}{=} 4 - \frac{9}{4} + 2$$
$$\frac{18 - 12 + 9}{4} \stackrel{?}{=} \frac{16 - 9 + 8}{4}$$
$$\frac{15}{4} = \frac{15}{4}.$$

If an equation involves multiplication by a constant, we first multiply and then combine any resulting like terms.

EXAMPLE 7.10   Solve $5(4 - 3t) - 2(5t - 2) = 49$.

Solution

First, we perform the multiplication and combine the resulting like terms:

$$5(4 - 3t) - 2(5t - 2) = 49$$
$$20 - 15t - 10t + 4 = 49$$
$$-25t + 24 = 49.$$

Then, we solve the resulting equation:

$$-25t + 24 - 24 = 49 - 24$$
$$-25t = 25$$
$$\frac{-25t}{-25} = \frac{25}{-25}$$
$$t = -1.$$

The solution is -1. To check, we substitute in the original equation. However, we do the operations inside the parentheses before multiplying by the constants:

$$5(4 - 3t) - 2(5t - 2) = 49$$
$$5[4 - 3(-1)] - 2[5(-1) - 2] \overset{?}{=} 49$$
$$5[4 + 3] - 2[-5 - 2] \overset{?}{=} 49$$
$$5[7] - 2[-7] \overset{?}{=} 49$$
$$35 + 14 \overset{?}{=} 49$$
$$49 = 49.$$

If an equation involves addition or subtraction of expressions, we first perform these operations and then combine any resulting like terms.

EXAMPLE 7.11   Solve $(4s - 9) - (8 - 3s) = 11$.

Solution

First, we perform the subtraction and combine the resulting like terms:

$$(4s - 9) - (8 - 3s) = 11$$
$$4s - 9 - 8 + 3s = 11$$
$$7s - 17 = 11.$$

Then, we solve the resulting equation:

$$7s - 17 + 17 = 11 + 17$$
$$7s = 28$$
$$\frac{7s}{7} = \frac{28}{7}$$
$$s = 4.$$

The solution is 4. To check, we substitute in the original equation. However, we do the operations inside the parentheses before subtracting the expressions:

$$(4s - 9) - (8 - 3s) = 11$$
$$[4(4) - 9] - [8 - 3(4)] \overset{?}{=} 11$$
$$[16 - 9] - [8 - 12] \overset{?}{=} 11$$
$$[7] - [-4] \overset{?}{=} 11$$
$$7 + 4 \overset{?}{=} 11$$
$$11 = 11.$$

EXAMPLE 7.12 Solve $2(x + 3) - (x - 2) = 5$.

Solution Simplifying the algebraic expression and then solving the resulting equation,

$$2(x + 3) - (x - 2) = 5$$
$$2x + 6 - x + 2 = 5$$
$$x + 8 = 5$$
$$x + 8 - 8 = 5 - 8$$
$$x = -3.$$

The solution is -3. You should check this solution. Be sure to substitute in the original equation and to do the operations inside the parentheses first.

EXAMPLE 7.13 Solve $4(y - 3) + y = 1 - (y + 13)$.

Solution Simplifying the algebraic expression on each side and then solving the resulting equation,

$$4(y - 3) + y = 1 - (y + 13)$$
$$4y - 12 + y = 1 - y - 13$$
$$5y - 12 = -y - 12$$
$$5y + y - 12 = -y + y - 12$$
$$6y - 12 = -12$$
$$6y - 12 + 12 = -12 + 12$$
$$6y = 0$$
$$\frac{6y}{6} = \frac{0}{6}$$
$$y = 0.$$

The solution is 0. You should check this solution. Be sure to substitute in the original equation and to do the operations inside the parentheses first.

In solving equations, there are two special cases which may be encountered. In one case the equation has no solution.

EXAMPLE 7.14 Solve $4 - (5 - 6x) = 3(2x - 3)$.

Solution   We simplify the algebraic expressions and attempt to solve the resulting equation:

$$4 - (5 - 6x) = 3(2x - 3)$$
$$4 - 5 + 6x = 6x - 9$$
$$-1 + 6x = 6x - 9.$$

Subtracting $6x$ from both sides of the resulting equation

$$-1 + 6x - 6x = 6x - 9 - 6x$$
$$-1 = -9.$$

Observe that the variable has disappeared. Moreover, the resulting equation is not true; that is, $-1 \neq -9$. This result tells us that there is no value of the variable which will cause the two sides to be equal. We say that the equation has **no solution.**

In the second special case, every possible value of the variable is a solution. For general linear equations, the possible values of the variable are all of the real numbers.

EXAMPLE 7.15   Solve $2(3x + 2) = 3(2x - 1) + 7$

Solution   We simplify the algebraic expressions and attempt to solve the resulting equation:

$$2(3x + 2) = 3(2x - 1) + 7$$
$$6x + 4 = 6x - 3 + 7$$
$$6x + 4 = 6x + 4.$$

Observe that the two sides of the resulting equation are identical. Therefore, the resulting equation is true for all values of x. Moreover, subtracting $6x$ from both sides gives us

$$6x + 4 - 6x = 6x + 4 - 6x$$
$$4 = 4.$$

The variable has disappeared and the resulting equation is true. This result tells us that every value of the variable will cause the two sides to be equal. We say that the solution is the set of **all real numbers.**

Exercise 7.2

Solve and check:

1. $6y - 2y - 1 = 5y - 2$

2. $4t - 2 - t = 3 + 2t - 1$

3. $2 - x + 5 = 3x - 4 + x$

4. $4 - 3z = 4z - 2 + z$

5. $2(x + 7) - 3(x + 6) = 4$

6. $2(4 - 3y) - 4(3y - 1) = 6$

7. $(2s - 3) - (9 - 4s) = -9$

8. $(4r - 5) - (6r - 8) = 7$

9. $4 - (3 - x) = 2(x - 6)$

10. $5(2z + 3) - (6z + 3) = 8$

11. $(4y + 9) - (2y + 6) = 3$

12. $3(2v - 1) - 5v = 5 - (2v + 8)$

13. $2(4r - 3) - 4 = 6 - (2 - 8r)$

14. $5(4 - s) = 4(s + 5) - 9s$

15. $4(3t + 2) = 3(4t - 2) + 14$

16. $10 - (6 - 2u) = 2(u - 1) + 5$

7.3   Equations with Constant Denominators

An equation may have denominators that are constants and still be a linear equation. For example, the equation

$$\frac{x}{5} = 2$$

is such an equation. Recall that we solve an equation of this type by multiplying each side by the nonzero number in the denominator:

$$\frac{x}{5} = 2$$
$$5\left(\frac{x}{5}\right) = 5(2)$$
$$x = 10.$$

If there is more than one term on one or both sides of the equation, we simplify the expressions after applying the multiplication rule.

EXAMPLE 7.16   Solve $\frac{x}{5} + 9 = 2x$.

Solution   We multiply both sides of the equation by 5:

$$\frac{x}{5} + 9 = 2x$$
$$5\left(\frac{x}{5} + 9\right) = 5(2x).$$

Now we simplify the expressions:

$$5\left(\frac{x}{5} + 9\right) = 5(2x)$$
$$5\left(\frac{x}{5}\right) + 5(9) = 5(2x)$$
$$x + 45 = 10x.$$

Then, completing the solution of the equation,

$$x + 45 = 10x$$
$$x + 45 - x = 10x - x$$
$$45 = 9x$$
$$\frac{45}{9} = \frac{9x}{9}$$
$$5 = x$$

or

$$x = 5.$$

The solution is 5. To check, we write

$$\frac{x}{5} + 9 = 2x$$

$$\frac{5}{5} + 9 \overset{?}{=} 2(5)$$

$$1 + 9 \overset{?}{=} 10$$

$$10 = 10.$$

The first step in the preceding example may be skipped if we observe that multiplying both sides of the equation by the constant denominator, and then simplifying, is the same as multiplying each term by the constant denominator. Thus, we may just multiply each term.

EXAMPLE 7.17   Solve $\dfrac{2u}{3} + 4 = \dfrac{u}{3}$.

Solution   We may multiply each term by 3:

$$\frac{2u}{3} + 4 = \frac{u}{3}$$

$$3\left(\frac{2u}{3}\right) + 3(4) = 3\left(\frac{u}{3}\right)$$

$$2u + 12 = u.$$

Then, completing the solution of the equation,

$$2u + 12 = u$$

$$2u + 12 - u = u - u$$

$$u + 12 = 0$$

$$u + 12 - 12 = 0 - 12$$

$$u = -12.$$

The solution is -12. You should check this solution by substituting in the original equation.

If there are two or more different constant denominators in an equation, we multiply each term by the least common denominator. Recall that the least common denominator is the smallest positive integer that is divisible by each of the denominators.

EXAMPLE 7.18   Solve $\dfrac{t}{2} - t = \dfrac{t}{6} + 1$.

Solution   The least common denominator is 6. Multiplying each term by 6,

$$\frac{t}{2} - t = \frac{t}{6} + 1$$

$$6\left(\frac{t}{2}\right) - 6(t) = 6\left(\frac{t}{6}\right) + 6(1)$$

$$3t - 6t = t + 6$$
$$-3t = t + 6$$
$$-3t - t = t + 6 - t$$
$$-4t = 6$$
$$\frac{-4t}{-4} = \frac{6}{-4}$$
$$t = -\frac{3}{2}.$$

The solution is $-\frac{3}{2}$. To check, we write

$$\frac{t}{2} - t = \frac{t}{6} + 1$$

$$\frac{-\frac{3}{2}}{2} - \left(-\frac{3}{2}\right) \overset{?}{=} \frac{-\frac{3}{2}}{6} + 1$$

$$\frac{-\frac{3}{2}}{\frac{2}{1}} + \frac{3}{2} \overset{?}{=} \frac{-\frac{3}{2}}{\frac{6}{1}} + 1$$

$$-\frac{3}{2} \cdot \frac{1}{2} + \frac{3}{2} \overset{?}{=} -\frac{3}{2} \cdot \frac{1}{6} + 1$$

$$-\frac{3}{4} + \frac{3}{2} \overset{?}{=} -\frac{1}{4} + 1$$

$$-\frac{3}{4} + \frac{6}{4} \overset{?}{=} -\frac{1}{4} + \frac{4}{4}$$

$$\frac{-3 + 6}{4} \overset{?}{=} \frac{-1 + 4}{4}$$

$$\frac{3}{4} = \frac{3}{4}.$$

In some equations a constant denominator applies to an entire expression.

EXAMPLE 7.19 Solve $\dfrac{3y - 1}{3} = \dfrac{5}{3}$.

Solution   The constant denominator 3 applies to the entire expression $3y - 1$. We multiply both sides of the equation by 3:

$$\frac{3y - 1}{3} = \frac{5}{3}$$

$$3\left(\frac{3y - 1}{3}\right) = 3\left(\frac{5}{3}\right).$$

Since the denominator 3 applies to the entire numerator $3y - 1$,

$$3\left(\frac{3y - 1}{3}\right) = 3y - 1.$$

Therefore,

$$3\left(\frac{3y - 1}{3}\right) = 3\left(\frac{5}{3}\right)$$
$$3y - 1 = 5$$
$$3y - 1 + 1 = 5 + 1$$
$$3y = 6$$
$$\frac{3y}{3} = \frac{6}{3}$$
$$y = 2.$$

The solution is 2. To check, we substitute in the original equation:

$$\frac{3y - 1}{3} = \frac{5}{3}$$
$$\frac{3(2) - 1}{3} \stackrel{?}{=} \frac{5}{3}$$
$$\frac{6 - 1}{3} \stackrel{?}{=} \frac{5}{3}$$
$$\frac{5}{3} = \frac{5}{3}.$$

Recall from Unit 1 that, for example, $\frac{x}{2}$ is the same as $\frac{1}{2}x$. Similarly, an expression such as $\frac{3y - 1}{3}$ is the same as $\frac{1}{3}(3y - 1)$. Thus, to solve an equation such as

$$\frac{1}{3}(3y - 1) = \frac{5}{3},$$

we may write

$$\frac{3y - 1}{3} = \frac{5}{3}$$

and continue as in the preceding example.

Exercise 7.3

Solve and check:

1. $\dfrac{x}{3} = 5$

2. $\dfrac{y}{-2} = 6$

3. $\dfrac{2u}{3} - 3 = 2$

4. $\dfrac{z}{3} + 8 = 3z$

5. $\dfrac{3y}{5} + 5 = \dfrac{-2}{5}$

6. $\dfrac{3t}{4} - 8 = \dfrac{5t}{4}$

7. $\dfrac{r}{2} + 5 = \dfrac{1}{3}$

8. $\dfrac{3s}{4} = \dfrac{2s}{5} + 1$

9. $\dfrac{5x}{6} - \dfrac{2}{3} = \dfrac{1}{2}$

10. $\dfrac{3z}{10} + \dfrac{1}{2} = \dfrac{z}{5}$

11. $\dfrac{t - 5}{2} = \dfrac{1}{2}$

12. $\dfrac{4s + 9}{4} = \dfrac{3}{2}$

13. $\dfrac{1}{2}(9v + 11) = 5$

14. $\dfrac{1}{5}(2u + 7) = u - 1$

15. $\dfrac{12y - 5}{6} = 2y - 1$

16. $\dfrac{4x - 9}{3} + 3 = \dfrac{x}{3} + x$

Self-test

Solve and check:

1. $6x - 2 = 2x - 9$

1. _____

2. $\dfrac{5y + 2}{5} = y - 2$

2. _____

3. $\dfrac{2t}{3} - 5 = -7$

3. _____

4. $3(s - 2) + 5(s + 1) = 11$

4. _____

5. $4 - (2z - 6) = -2(z - 5)$

5. _____

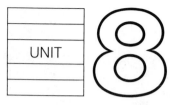

UNIT **8**

Applications

INTRODUCTION

Solving equations is a means of solving applied problems. An applied problem is a verbal description of a situation that can be described in mathematical terms. In this unit you will learn how to translate the words of an applied problem into mathematical terms, and how to form an equation. You can then solve the equation to find the solution of the problem.

OBJECTIVES

When you have finished this unit you should be able to:

1. Translate verbal statements into mathematical expressions, and use such expressions to solve problems describing numbers.
2. Solve problems describing angles of a triangle, and perimeters of triangles and rectangles.
3. Solve problems involving values and mixtures.
4. Solve problems involving ratio and proportion.

8.1 Translating Phrases

Solving an applied problem is very much like translating a paragraph from one language to another. We must translate a paragraph written in English into another language called "mathematics." This process involves learning the meaning of English words in terms of mathematical symbols, and then translating English phrases into mathematical phrases. We start with two basic translations.

"More" in English means "plus," the addition symbol, in mathematics. For example, if one number is four more than another number, we must add 4 to the smaller number in order to represent the larger number. If x represents the smaller number, then $x + 4$ represents the larger number.

EXAMPLE 8.1 Translate into a mathematical expression: "Ten more than a number."

Solution Suppose we use x to represent the number. Then, "10 more than x" is written

$$x + 10.$$

"Less" in English means "minus," the subtraction symbol, in mathematics. For example, if one number is six less than another number, we must subtract 6 from the larger number in order to represent the smaller number. If x represents the larger number, then $x - 6$ represents the smaller number.

EXAMPLE 8.2 Translate into a mathematical expression: "Fifteen less than a number."

Solution Suppose we use x to represent the number. Then "15 less than x" is written as 15 subtracted from x, or

$$x - 15.$$

☐ **EXAMPLE 8.3**   Translate into a mathematical expression: "Fifteen less a number."

Solution   Suppose we use x to represent the number. Then "15 less x" is written as x subtracted from 15, or

$$15 - x.$$

It is a common error to confuse "less than" and "less." Observe that 15 *less than* x is $x - 15$, whereas 15 *less* x is $15 - x$.

Subtraction also can be used to represent one of two numbers when their sum is given.

☐ **EXAMPLE 8.4**   Translate into a mathematical expression: "A number which, with another number x, makes a sum of 50."

Solution   One number is given to be x. Since x and the other number together make 50, the other number is 50 less x, or

$$50 - x.$$

We observe that $(50 - x) + x$ makes a sum of 50.

We now introduce a third important translation. "Of" in English means "times," the multiplication symbol, in mathematics. For example, if one number is x, three-fourths of that number is $\frac{3}{4}x$.

☐ **EXAMPLE 8.5**   Translate into a mathematical expression: "One-fifth of a number."

Solution   Suppose the number is x. Then, one-fifth of x is written

$$\frac{1}{5}x.$$

We will often find "of" combined with "more" or "less" in one phrase. In this case, we need another symbol of mathematics, the parentheses. Parentheses tell us which comes first, the "of" or the "more" or "less." For example, consider the phrases "one-half of two more than a number" and "two more than one-half of a number." For the first, we must write the translation of "two more than a number," and then translate "one-half of" that entire quantity. We write $x + 2$, and then $\frac{1}{2}(x + 2)$, using parentheses to indicate one-half of the entire quantity.

For the second phrase, "two more than one-half of a number," we must write the translation of "one-half of a number," and then translate "two more than" that quantity. We write $\frac{1}{2}x$, and then $\frac{1}{2}x + 2$. In this case there are no parentheses because we have one-half of x only, not of a sum or difference.

☐ **EXAMPLE 8.6**   Translate into mathematical expressions:

a. Three-halves of six less than a number.

b. Six less than three-halves of a number.

Solutions   Suppose in each case the number is x.

a. We have six less than x, or $x - 6$. Then, we have three-halves of that entire quantity. The translation is

$$\frac{3}{2}(x - 6).$$

b. We have three-halves of x only, or $\frac{3}{2}x$. Then, we have six less than that quantity. The translation is

$$\frac{3}{2}x - 6.$$

Recall that an equation consists of two algebraic expressions with an equal sign between them. The English word "is" represents the equal sign.

EXAMPLE 8.7 Three times a number is six more than the number. Find the number.

Solution Suppose the number is x. Three times the number is represented by the algebraic expression

$$3x.$$

Six more than the number is represented by the algebraic expression

$$x + 6.$$

"Is" translates to an equal sign. We translate the sentence into a mathematical sentence:

Three times a number is six more than the number.
$$3x \qquad = \qquad x + 6.$$

We have the equation

$$3x = x + 6.$$

Solving this equation,

$$2x = 6$$
$$x = 3.$$

The number is 3. To check, three times the number 3 is 9, and six more than the number 3 is also 9. Therefore, the two sides both represent 9 and are equal. You must not use the equation to check since you may have translated the words incorrectly. You must use the original wording of the problem.

Again, when "of" is combined with "more" or "less," we must determine whether we need parentheses.

EXAMPLE 8.8 One-third of nine less than a number is five. Find the number.

Solution Suppose the number is x. We have nine less than the number, or

$$x - 9.$$

Then, we have one-third of the entire quantity "nine less than the number," or

$$\frac{1}{3}(x - 9).$$

The equation is

$$\frac{1}{3}(x - 9) = 5.$$

To solve this equation, we multiply both sides by 3:

$$3\left(\frac{1}{3}\right)(x - 9) = 3(5)$$
$$x - 9 = 15$$
$$x = 24.$$

The number is 24. To check, we observe that nine less than 24 is 15 and then one-third of 15 is 5.

Exercise 8.1

Translate into mathematical expressions:

1. Fourteen more than a number x.

2. A number that is three more than x.

3. Five less than a number x.

4. A number that is 30 less than x.

5. Seven less a number x.

6. A number x less seven.

7. A number that with x makes a sum of nine.

8. A number that with x makes a sum of 200.

9. Three of a number x.

10. Ten of a number x.

11. One-third of a number x.

12. Five-halves of a number x.

13. Eighteen more than two-thirds of a number x.

14. Ten less than one-fifth of a number x.

15. Two-fifths of eight more than a number x.

16. Three-fourths of twelve less than a number x.

17. Twice a number is eight more than the number. Find the number.

18. Eleven times a number is five more than the number. Find the number.

19. One-half of a number is three less than the number. Find the number.

20. One-fifth of a number is twelve more than the number. Find the number.

21. One-fourth of five less than a number is six. Find the number.

22. One-fifth of seven more than a number is twelve. Find the number.

23. Five less than one-fourth of a number is three-fourths. Find the number.

24. Five more than two-thirds of a number is five-thirds. Find the number.

8.2   Triangles and Rectangles

One application of translating phrases involves the angles of a **triangle.** You have probably learned in previous courses that the sum of the angles of a triangle is 180° (180 degrees). This means that, if we add the degree measures of each of the three angles of any triangle, the sum is 180°. This fact is proved in Euclidean geometry.

EXAMPLE 8.9   If two angles of a triangle are 52° and 73°, how large is the third angle?

Solution   Suppose x is the size of the third angle. We draw a diagram of the triangle:

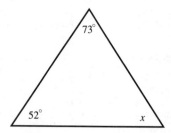

The sum of the three angles of the triangle is 180°. Therefore, we have the equation

$$52 + 73 + x = 180.$$

Simplifying and solving this equation,

$$125 + x = 180$$
$$x = 55.$$

The third angle is 55°. To check, we observe that

$$52 + 73 + 55 = 180.$$

In translating phrases, we used mathematical translations of phrases involving the words "more" and "less." We use the same translations to solve problems describing the angles of a triangle.

EXAMPLE 8.10   In a triangle, the second angle is 15° less than the first, and the third angle is 24° more than the first. Find the size of each angle of the triangle.

Solution   Suppose x is the size of the first angle. Then the second angle is 15° less. Therefore, the second angle is

$$x - 15.$$

The third angle is 24° more than the first angle. Therefore, the third angle is

$$x + 24.$$

Thus, the three angles are x, $x - 15$, and $x + 24$. We draw a diagram of the triangle:

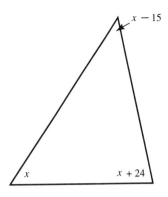

Since the sum of the angles is 180°,

$$x + (x - 15) + (x + 24) = 180.$$

We remove the parentheses, combine like terms, and solve the equation:

$$x + x - 15 + x + 24 = 180$$
$$3x + 9 = 180$$
$$3x = 171$$
$$x = 57.$$

The first angle is 57°. To find the other two angles, we calculate

$$x - 15 = 57 - 15 = 42$$

and

$$x + 24 = 57 + 24 = 81.$$

The angles are 57°, 42°, and 81°. To check, observe that

$$57 + 42 + 81 = 180.$$

EXAMPLE 8.11   In a triangle, the largest angle is three times the size of the smallest angle. The third angle is 5° more than the smallest angle. Find the size of each angle of the triangle.

Solution   Suppose x is the size of the smallest angle. Then three times the size of the smallest angle is

$$3x,$$

and 5° more than the smallest angle is

$$x + 5.$$

A diagram of the triangle is

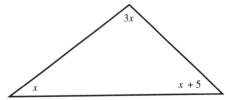

Therefore,

$$x + 3x + (x + 5) = 180$$
$$x + 3x + x + 5 = 180$$
$$5x + 5 = 180$$
$$5x = 175$$
$$x = 35.$$

The smallest angle is 35°. The largest angle is

$$3x = 3(35) = 105,$$

and the third angle is

$$x + 5 = 35 + 5 = 40.$$

The angles are 35°, 105°, and 40°. To check, observe that the sum of these angles is 180°.

The angles described may not always be compared to the angle that is mentioned first.

EXAMPLE 8.12   In a triangle, the third angle is one-third of the second angle. The first angle is 16° less than the second angle. Find the size of each angle of the triangle.

Solution

In this example, the first and third angles are compared to the second. Therefore, we let x be the size of the second angle. Then, the third angle is

$$\frac{1}{3}x$$

and the first angle is

$$x - 16:$$

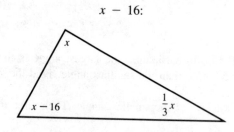

Therefore,

$$x + \frac{1}{3}x + (x - 16) = 180$$

$$x + \frac{1}{3}x + x - 16 = 180$$

$$\frac{7}{3}x - 16 = 180$$

$$\frac{7}{3}x = 196$$

$$x = \frac{3}{7}(196)$$

$$x = 84.$$

The second angle is 84°. The third angle is

$$\frac{1}{3}x = \frac{1}{3}(84) = 28$$

and the first angle is

$$x - 16 = 84 - 16 = 68.$$

The angles are 68°, 84°, and 28°. The sum of these angles is 180°.

EXAMPLE 8.13   In a triangle, the third angle is twice the second angle, and the second angle is 40° more than the first angle. Find the size of each angle of the triangle.

Solution   We let x be the size of the first angle. Then, the second angle is

$$x + 40.$$

The third angle is twice the second. Since the second angle is $x + 40$, the third angle is

$$2(x + 40):$$

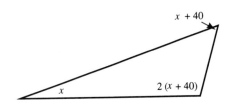

Therefore,

$$x + (x + 40) + 2(x + 40) = 180$$
$$x + x + 40 + 2x + 80 = 180$$
$$4x + 120 = 180$$
$$4x = 60$$
$$x = 15.$$

The first angle is 15°. Then, the second angle is

$$x + 40 = 15 + 40 = 55$$

and the third angle is

$$2(x + 40) = 2(55) = 110.$$

The angles are 15°, 55°, and 110°. You should check this solution.

A triangle is an example of a **polygon.** The **perimeter** of a polygon is the sum of the lengths of its sides. If we are given the perimeter of a triangle and information about its sides to construct algebraic expressions, we can find the lengths of the sides.

EXAMPLE 8.14   A triangle has a perimeter of 54 inches. One side is 3 inches more than the shortest side. The longest side is 12 inches more than the shortest side. Find the sides of the triangle.

Solution   Suppose x is the length of the shortest side. Then, the next side is 3 more, that is,

$$x + 3.$$

The longest side is 12 more than the shortest side, or

$$x + 12.$$

We draw a diagram of the triangle:

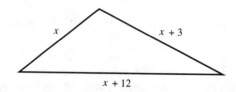

Since the sum of the sides is 54,

$$x + (x + 3) + (x + 12) = 54$$
$$x + x + 3 + x + 12 = 54$$
$$3x + 15 = 54$$
$$3x = 39$$
$$x = 13.$$

The shortest side is 13 inches. Then, the next side is

$$x + 3 = 16,$$

and the longest side is

$$x + 12 = 25.$$

The sides are 13 inches, 16 inches, and 25 inches. To check,

$$13 + 16 + 25 = 54.$$

EXAMPLE 8.15   One side of a triangle is twice the second side. The third side is 30 centimeters more than the second side. If the perimeter is 130 centimeters, find the sides of the triangle.

Solution   We let x represent the second side. Then the first side is

$$2x,$$

and the third side is

$$x + 30.$$

A diagram of the triangle is

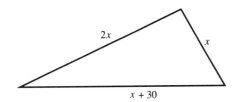

The sum of the sides is

$$
\begin{aligned}
x + 2x + (x + 30) &= 130 \\
x + 2x + x + 30 &= 130 \\
4x + 30 &= 130 \\
4x &= 100 \\
x &= 25.
\end{aligned}
$$

The second side is 25 centimeters. The first side is

$$2x = 2(25) = 50$$

and the third side is

$$x + 30 = 25 + 30 = 55.$$

The sides are 25 centimeters, 50 centimeters, and 55 centimeters. To check, observe that the sum of the sides is

$$25 + 50 + 55 = 130.$$

Another common polygon is the **rectangle:**

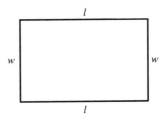

The two sides marked w, for **width,** are equal, and the two sides marked l, for **length,** are equal. All of the angles are 90°. The perimeter is

$$w + l + w + l = 2w + 2l.$$

EXAMPLE 8.16   The width of a rectangle is 6 feet less than the length. The perimeter is 32 feet. Find the length and the width.

Solution   If the length is x, then the width is $x - 6$:

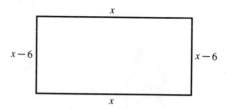

The perimeter of a rectangle is $2w + 2l$; therefore,

$$2(x - 6) + 2x = 32$$
$$2x - 12 + 2x = 32$$
$$4x - 12 = 32$$
$$4x = 44$$
$$x = 11.$$

The length is 11 feet, and the width is $x - 6 = 11 - 6 = 5$ feet. To check, the perimeter is

$$2(5) + 2(11) = 10 + 22$$
$$= 32.$$

Exercise 8.2

1. If two angles of a triangle are 31° and 65°, how large is the third angle?

2. If two angles of a triangle are 45° and 23°, how large is the third angle?

3. In a triangle, the second angle is 10° more than the first. The third angle is 4° less than the first. Find the size of each angle of the triangle.

4. In a triangle, the second angle is 20° less than the first, and the third angle is 31° less than the first. Find the size of each angle of the triangle.

5. In a triangle, the second angle is 16° less than the first. The third angle is twice the size of the first angle. Find the size of each angle of the triangle.

6. In a triangle, the second angle is twice the size of the first, and the third angle is three times the size of the first angle. Find the size of each angle of the triangle.

7. In a triangle, the second angle is two-thirds of the first. The third angle is 4° more than the first angle. Find the size of each angle of the triangle.

8. In a triangle, the second angle is three-quarters of the third angle, and the first angle is one-half of the third angle. Find the size of each angle of the triangle.

9. In a triangle, the third angle is four-thirds of the second angle, and the second angle is three times the first angle. Find the size of each angle of the triangle.

10. In a triangle, the second angle is 12° more than the first angle, and the third angle is twice the second angle. Find the size of each angle of the triangle.

11. In a triangle, the second side is 8 feet more than the shortest side, and the longest side is 12 feet more than the shortest side. The perimeter is 56 feet. Find the sides of the triangle.

12. The perimeter of a triangle is 33 meters. The shortest side is 3 meters less than the second side, and the longest side is 3 meters more than the second side. Find the sides of the triangle.

13. The perimeter of a triangle is 35 yards. The longest side is 3 times the shortest side, and the other side is 10 yards more than the shortest side. Find the sides of the triangle.

14. One side of a triangle is twice the second side. The second side is 4 inches less than the third side. The perimeter is 76 inches. Find the sides of the triangle.

15. The width of a rectangle is 10 feet less than the length. The perimeter is 100 feet. Find the length and the width.

16. The length of a rectangle is 22 centimeters more than the width. The perimeter is 136 centimeters. Find the width and the length.

17. The width of a rectangle is one-half of the length. The perimeter is 48 inches. Find the length and the width.

18. The width of a rectangle is two-thirds of the length. The perimeter is $6\frac{2}{3}$ yards. Find the length and the width.

19. The length of a rectangle is five-fourths of the width. The perimeter is 45 feet. Find the width and the length.

20. The length of a rectangle is twice the width. The perimeter is 16 meters. Find the width and the length.

8.3   Values and Mixtures

A category of applications describes the value of a collection of two or more types of things. For example, the collection might consist of pennies and nickels. Such a collection has a specific value. The value depends on how many pennies there are, and how many nickels there are.

EXAMPLE 8.17   A collection of pennies and nickels has a value of 83¢. If there are 27 coins in the collection, how many pennies and how many nickels are there?

Solution   Problems involving values often are easiest to solve using a chart. We draw a grid like this:

Going down the left-hand side of the grid are spaces for the two types of things, in this example pennies and nickels, and for the value of the collection:

Pennies		
Nickels		
Collection		

Across the top of the grid are spaces for the value of each thing, the number of each, and the total value of each:

	Values	Number of coins	Total value
Pennies			
Nickels			
Collection			

We begin by filling in the parts of the chart that we know. The value of a penny is 1¢ and the value of a nickel is 5¢. Also, the number of coins in the collection is 27, and the total value of the collection is 83¢:

	Values	Number of coins	Total value
Pennies	1		
Nickels	5		
Collection		27	83

Now, suppose x is the number of pennies. Then the number of nickels is the number that with x makes a sum of 27, or $27 - x$. We fill in this information on the chart:

	Values	Number of coins	Total value
Pennies	1	x	
Nickels	5	$27 - x$	
Collection		27	83

Finally, each penny is worth 1¢, and there are x pennies; therefore, the total value of the pennies is $1(x)$. Each nickel is worth 5¢ and there are $27 - x$ nickels; therefore, the total value of the nickels is $5(27 - x)$. We complete the chart:

	Value	Number of coins	Total value
Pennies	1	x	$1(x)$
Nickels	5	$27 - x$	$5(27 - x)$
Collection		27	83

The total value of the pennies plus the total value of the nickels is equal to the total value of the collection. Using the last column of the chart, we have the equation

$$1(x) + 5(27 - x) = 83.$$

Simplifying, and solving for x,

$$x + 135 - 5x = 83$$
$$-4x + 135 = 83$$
$$-4x = -52$$
$$x = 13.$$

There are 13 pennies in the collection. Therefore, since

$$27 - x = 27 - 13 = 14,$$

there are 14 nickels in the collection. To check, 13 pennies are worth 13¢, and 14 nickels are worth 5(14), or 70¢. Therefore, the total value is 83¢.

EXAMPLE 8.18   A collection of quarters and dimes has a value of $10.25. If there are 50 coins in the collection, how many quarters and how many dimes are there?

Solution   We make a chart as in the preceding example. However, observe that, if we write the value of each coin in cents, we must write the total value in cents. Thus, we write the total value of $10.25 as 1025¢:

	Value	Number of coins	Total value
Quarters	25	x	25(x)
Dimes	10	50 − x	10(50 − x)
Collection		50	1025

The equation is

$$25(x) + 10(50 - x) = 1025$$
$$25x + 500 - 10x = 1025$$
$$15x + 500 = 1025$$
$$15x = 525$$
$$x = 35.$$

There are 35 quarters. Therefore, since

$$50 - x = 50 - 35 = 15,$$

there are 15 dimes. You should check that these numbers of coins give the correct total value.

EXAMPLE 8.19   A collection of nickels and quarters has 15 more nickels than quarters. The total value of the collection is $1.65. How many nickels and how many quarters are there?

Solution   Observe that we are not given a total number of coins. Instead, the number of quarters is related to the number of nickels. If x is the number of quarters, then $x + 15$ is the number of nickels. We use a chart similar to those in the preceding examples:

	Value	Number of coins	Total value
Nickels	5	$x + 15$	$5(x + 15)$
Quarters	25	x	$25(x)$
Collection			165

The equation is

$$5(x + 15) + 25x = 165$$
$$5x + 75 + 25x = 165$$
$$30x + 75 = 165$$
$$30x = 90$$
$$x = 3.$$

Therefore, there are 3 quarters. Since

$$x + 15 = 3 + 15 = 18,$$

there are 18 nickels. You should check that these numbers of coins give the correct total value.

If we have more than two types of things in the collection, we need extra lines in the chart. If there are three types of things, we add one more line to the chart.

EXAMPLE 8.20   A collection of nickels, dimes, and quarters has a value of $3.25. There are four more dimes than nickels, and three less quarters than nickels. How many of each type of coin are there?

Solution   The chart is

	Value	Number of coins	Total value
Nickels	5	x	$5(x)$
Dimes	10	$x + 4$	$10(x + 4)$
Quarters	25	$x - 3$	$25(x - 3)$
Collection			325

The equation is

$$5(x) + 10(x + 4) + 25(x - 3) = 325$$
$$5x + 10x + 40 + 25x - 75 = 325$$
$$40x - 35 = 325$$

$$40x = 360$$
$$x = 9.$$

Also,

$$x + 4 = 9 + 4 = 13$$

and

$$x - 3 = 9 - 3 = 6.$$

Therefore, there are 9 nickels, 13 dimes, and 6 quarters. To check, we find the value of the nickels to be $0.45, the value of the dimes $1.30, and the value of the quarters $1.50. These numbers of coins give

$$\$0.45 + \$1.30 + \$1.50 = \$3.25.$$

The values involved in this category of applications may be values of things other than coins. In such applications, the values must be stated in the problem.

EXAMPLE 8.21   A total of 120 hot dogs and hamburgers are sold for $92.40. If the hot dogs are sold for 60¢ each and the hamburgers are sold for 90¢ each, how many hot dogs and how many hamburgers are sold?

Solution   We use a chart similar to those in the preceding examples, filling in the given values:

	Value	Number of items	Total value
Hot dogs	60	x	$60(x)$
Hamburgers	90	$120 - x$	$90(120 - x)$
Collection		120	9240

The equation is

$$60(x) + 90(120 - x) = 9240$$
$$60x + 10{,}800 - 90x = 9240$$
$$-30x + 10{,}800 = 9240$$
$$-30x = -1560$$
$$x = 52,$$

and

$$120 - x = 120 - 52 = 68.$$

Therefore, 52 hot dogs and 68 hamburgers are sold. To check, remember that you must always use the wording of the problem and not your equation—the equation itself could be wrong. Using the wording of the problem, if 52 hot dogs are sold at 60¢ each, the value is $31.20, and if 68 hamburgers are sold at 90¢ each, the value is $61.20. Therefore, the total value is $92.40.

Mixture problems are of a similar type. For mixture problems, we need a chart with lines for each component of the mixture and for the mixture itself, and spaces for the unit price of each component, the numbers of units, and the total prices:

	Unit price	Number of units	Total price
First component			
Second component			
Mixture			

EXAMPLE 8.22   A 10-pound bag of potatoes contains Idahos, which sell for 45¢ a pound, and all-purpose Maines, which sell for 33¢ a pound. The bag costs $3.66. How many pounds of each kind of potato does it contain?

Solution   We make a chart of the price per pound and number of pounds of each type of potato. If there are x pounds of Idahos, and a total of 10 pounds in the mixture, then there are $10 - x$ pounds of Maine potatoes:

	Unit price	Number of units	Total price
Idaho	45	x	
Maine	33	$10 - x$	
Mixture		10	

The total price of the Idahos is 45¢ a pound times x pounds, or $45x$. The total price of the Maines is 33¢ a pound times $10 - x$ pounds, or $33(10 - x)$. We are not given the unit price of the mixture, or bag, but the total price of the bag is given as $3.66, or 366¢. Now we can fill in the total prices on our chart:

	Unit price	Number of units	Total price
Idaho	45	x	$45x$
Maine	33	$10 - x$	$33(10 - x)$
Mixture		10	366

The sum of the total prices of each type of potato is equal to the total price of the bag. Thus, the equation is

$$45x + 33(10 - x) = 366.$$

Solving this equation,

$$45x + 330 - 33x = 366$$
$$12x + 330 = 366$$

$$12x = 36$$
$$x = 3.$$

The bag contains three pounds of Idaho potatoes and seven pounds of Maine potatoes. To check, we must use the wording of the problem, not the equation. Three pounds of Idaho potatoes at 45¢ a pound cost $1.35. Seven pounds of Maine potatoes at 33¢ a pound cost $2.31. The total price is $1.35 + $2.31 = $3.66 for the bag of potatoes.

In the preceding value and mixture problems, we were given the total value or price of the collection or mixture. In mixture problems we may be given the unit price of the mixture instead, for example, the price per pound. In this case, we put the unit price in the first space of the mixture line, and multiply by the number of units to get the total price.

EXAMPLE 8.23 Creme-filled chocolates worth $2.50 a pound are mixed with plain chocolates worth $1.30 a pound to make 20 pounds of a mixture which sells for $1.60 a pound. How many pounds of each type of chocolate are in the mixture?

Solution We make a chart for the price per pound and number of pounds of each type of chocolate. Since the total price of the mixture is not given, we use the final line to give the price per unit and number of units for the mixture:

	Unit price	Number of units	Total price
Creme	2.50	x	$2.50x$
Plain	1.30	$20 - x$	$1.30(20 - x)$
Mixture	1.60	20	32.00

Since the sum of the total price of each type of chocolate is equal to the total price for the mixture, the equation is:

$$2.50x + 1.30(20 - x) = 32.00.$$

It is often easiest to write the prices in cents before solving the equation:

$$250x + 130(20 - x) = 3200$$
$$250x + 2600 - 130x = 3200$$
$$120x = 600$$
$$x = 5.$$

There are 5 pounds of creme-filled chocolates and 15 pounds of plain chocolates in the mixture. You should check these answers, using the original wording of the problem.

Exercise 8.3

1. A collection of pennies and dimes has a value of 73¢. If there are 28 coins in the collection, how many pennies and how many dimes are there?

2. A collection of 20 pennies and nickels has a value of 64¢. How many pennies and how many nickels are there?

3. A collection of nickels and dimes has a value of $2.90. If there are 40 coins in the collection, how many nickels and how many dimes are there?

4. A collection of 36 dimes and quarters has a value of $6.15. How many dimes and how many quarters are there in the collection?

5. A collection of quarters and half dollars has a value of $14.50. There are 7 more quarters than half dollars. How many quarters and how many half dollars are there?

6. A collection of nickels and quarters has 12 less nickels than quarters. If the total value of the collection is $4.20, how many nickels and how many quarters are there?

7. A collection of pennies, nickels, and dimes has a value of $5.80. There are 6 less nickels than pennies and 11 less dimes than pennies. How many of each type of coin are there?

8. A collection of nickels, dimes, and quarters has twice as many nickels as quarters. There are 22 more dimes than quarters. If the total value of the collection is $10.75, how many of each type of coin are there?

9. A total of 190 theatre tickets are sold for $2300.00. If tickets for the closer rows are sold for $15.00 each and tickets for the back rows are sold for $10.00 each, how many of each are sold?

10. A total of 472 tickets to a dance are sold for $1050.00. If student tickets are sold for $2.00 each and guest tickets are sold for $2.50 each, how many of each are sold?

11. Snacks of potato chips and pretzels are sold for a total of $18.60. The potato chips are sold for 40¢ a package and the pretzels are sold for 35¢ a package. If six more packages of pretzels than potato chips are sold, how many packages of each are sold?

12. Coffee and tea are sold for a total of $39.00. Coffee is sold for 50¢ a cup and tea is sold for 40¢ a cup. If 24 more cups of coffee than tea are sold, how many cups of each are sold?

13. Two varieties of pears are mixed in a 5-pound package. The package contains Bosc pears, which sell for 89¢ a pound, and Bartlett pears, which sell for 69¢ a pound. The package sells for $3.85. How many pounds of each variety of pear does it contain?

14. A 50-pound bag of sand sells for $4.25. The bag contains a mixture of fine sand worth 12¢ a pound and coarse sand worth 5¢ a pound. How many pounds of each is in the bag?

15. An assortment of 30 pens contains two types. There are ball-point pens, which sell for 45¢ each, and felt-tip pens, which sell for 95¢ each. The assortment sells for $18.50. How many pens of each type does it contain?

16. A mixed case of 24 cans includes light tuna worth 90¢ a can and white tuna worth $1.30 a can. The case sells for $25.20. How many cans of each type does it contain?

17. Ten pounds of bean salad are made by mixing kidney beans worth 60¢ a pound and cut green beans worth $1.20 a pound. The mixture is worth 96¢ a pound. How many pounds of each type of bean does it contain?

18. Beef flavor and liver flavor dog food are mixed to make an 80-ounce bag. The beef flavor costs 20¢ an ounce and the liver flavor costs 30¢ an ounce. The bag is worth $22\frac{1}{2}$¢ an ounce. How many ounces of each flavor are in it?

19. A mixture of 25 pecks of soil is made to sell for $1.26 a peck. It contains peat moss worth $1.40 a peck and potting soil worth $1.20 a peck. How many pecks of each does it contain?

20. Olive oil and vegetable oil are mixed to make 50 quarts of salad oil worth $2.50 a quart. The olive oil is worth $3.60 a quart and the vegetable oil is worth $1.60 a quart. How many quarts of each does the salad oil contain?

8.4   Ratio and Proportion

A **ratio** is a fraction. For example, $\frac{3}{7}$ is the ratio of three to seven. This ratio is also written

$$3:7.$$

A ratio need not involve only integers. The ratio

$$3\frac{1}{2}:2$$

is read "three and one-half to two," and is the fraction

$$\frac{3\frac{1}{2}}{2}.$$

A **proportion** consists of two equal ratios. For example,

$$\frac{3}{7} = \frac{9}{21}$$

is a proportion. It can be read "three is to seven as nine is to twenty-one."

Now, consider the proportion

$$\frac{a}{b} = \frac{c}{d},$$

where a, b, c, and d may be constants or variables. We know that we may multiply both sides of an equation by the same number. Suppose we multiply each side of this proportion by bd:

$$\frac{a}{b} = \frac{c}{d}$$

$$bd\left(\frac{a}{b}\right) = bd\left(\frac{c}{d}\right)$$

$$da = bc$$

or

$$ad = bc.$$

This results tells us that the product of the numerator of the first ratio and the denominator of the second is equal to the product of the denominator of the first ratio and the numerator of the second. We use the pattern

$$\frac{a}{b} \diagup\!\!\!\!\diagdown \frac{c}{d}$$

$$ad = bc,$$

where the arrows show which parts to multiply.

EXAMPLE 8.24   A number is to 84 as 11 is to 21. Find the number.

Solution   Let x be the number. The proportion is

$$\frac{x}{84} = \frac{11}{21}.$$

Then,

$$\frac{x}{84} \diagdown \frac{11}{21}$$

$$x(21) = 84(11)$$

$$x = \frac{84(11)}{21}$$

$$x = 44.$$

The number is 44. To check, observe that

$$\frac{44}{84} = \frac{11(4)}{21(4)} = \frac{11}{21},$$

and so the ratios are equal.

Ratios and proportions may be used to solve a variety of types of applied problems.

EXAMPLE 8.25   A car can go 22 miles on one gallon of gas. How much gas is needed for the car to go 440 miles?

Solution   Let x be the amount of gas. The first ratio is

22 miles : 1 gallon

or $\frac{22}{1}$. The second ratio is

440 miles : x gallons

or $\frac{440}{x}$. Observe that the ratios must go in the same order, that is, miles to gallons. The proportion is

$$\frac{22}{1} = \frac{440}{x}$$

$$\frac{22}{1} \diagdown \frac{440}{x}$$

$$22x = 1(440)$$

$$x = \frac{440}{22}$$

$$x = 20.$$

The amount of gas needed is 20 gallons. To check,

$$\frac{440}{20} = \frac{22(20)}{1(20)} = \frac{22}{1}.$$

EXAMPLE 8.26   On a map, a scale of 1 inch:16 miles is used. How many inches are needed to represent 200 miles?

Solution   If x is the number of inches, then the second ratio is

$$x \text{ inches}:200 \text{ miles},$$

where both ratios go in the order inches to miles. The proportion is

$$\frac{1}{16} = \frac{x}{200}$$

$$\frac{1}{16} \times \frac{x}{200}$$

$$1(200) = 16x$$

$$\frac{200}{16} = x$$

$$12.5 = x.$$

The number of inches is 12.5 or $12\frac{1}{2}$ inches. To check,

$$\frac{12.5}{200} = \frac{125}{2000} = \frac{1(125)}{16(125)} = \frac{1}{16},$$

or

$$\frac{12\frac{1}{2}}{200} = \frac{\frac{25}{2}}{\frac{200}{1}} = \frac{25}{2} \cdot \frac{1}{200} = \frac{25}{400} = \frac{1(25)}{16(25)} = \frac{1}{16}.$$

Ratios and proportions are useful in converting among measurement systems. Given the appropriate ratios, we can use a proportion to convert English measurements to metric measurements.

EXAMPLE 8.27   The ratio of inches to centimeters is approximately 1 inch:2.54 centimeters. How many centimeters are there in a foot?

Solution   A foot is 12 inches. Therefore, the second ratio is

$$12 \text{ inches}:x \text{ centimeters}.$$

The proportion is

$$\frac{1}{2.54} = \frac{12}{x}$$

$$\frac{1}{2.54} \times \frac{12}{x}$$

$$1x = 2.54(12)$$

$$x = 30.48.$$

There are 30.48 centimeters in a foot. To check, we compute each of the two ratios:

$$\frac{1}{2.54} \approx 0.3937$$

and

$$\frac{12}{30.48} \approx 0.3937,$$

where "≈" means "is approximately equal to." The ratios are equal to at least four decimal places.

 Exercise 8.4

1. A number is to 360 as 4 is to 9. Find the number.

2. Twelve is to a number as 2 is to 15. Find the number.

3. A car can go 18 miles on one gallon of gas. How much gas is needed for the car to go 90 miles?

4. A car can go 85 miles on 5 gallons of gas. How far can it go on 100 gallons of gas?

5. On a map, a scale of $\frac{1}{2}$ inch:3 miles is used. How many inches are needed to represent 24 miles?

6. On a map, a scale of 2 inches:25 miles is used. How many miles are represented by 7 inches?

7. The ratio of the height of a post to the length of its shadow is 4 feet:$2\frac{1}{2}$ feet. If your shadow is $3\frac{3}{4}$ feet long, how tall are you?

8. The ratio of your weight on the earth to your weight on the moon is 6 pounds:1 pound. If you weigh 135 pounds on the earth, how much do you weigh on the moon?

9. The ratio of inches to centimeters is approximately 1 inch:2.54 centimeters. How many centimeters are there in 16 inches?

10. The ratio of inches to centimeters is approximately 1 inch:2.54 centimeters. How many inches are there in 1 meter? (1 meter = 100 centimeters. Observe that 1 meter is a little more than a yard.)

11. The ratio of miles to kilometers is approximately 0.6 miles:1 kilometer. If you have to travel 90 kilometers, how many miles must you travel?

12. The ratio of quarts to liters is approximately 1.06 quarts:1 liter. How many quarts are there in 1.75 liters? (Observe that 1.75 liters is slightly less than 2 quarts, or a half gallon.)

Self-test

1. Translate into an algebraic expression:
   Five less than four-fifths of a number.

   1. _____

2. The ratio of the height of a tree to the length of its shadow is 5:2. If the length of the shadow is 9 meters, how tall is the tree?

   2. _____

3. You have dimes and nickels in your pocket worth a total of $1.45. If there are 20 coins, how many dimes and how many nickels do you have?

   3. _____

4. Ginger ale and juice concentrate are mixed to make 15 liters of fruit punch. The ginger ale is worth 60¢ a liter and the fruit concentrate is worth $1.50 a liter. If the mixture sells for 75¢ a liter, how many liters of ginger ale and how many liters of fruit concentrate does it contain?

   4. _____

5. The perimeter of a rectangle is 14 feet. The length is 2 feet more than the width. Find the width and the length.

   5. _____

UNIT **9**

Formulas

INTRODUCTION

In the preceding units you have learned to write and solve equations in one variable. A formula is a type of equation that relates two or more variables. In this unit you will learn how to evaluate formulas and how to solve formulas. You will also learn to write and solve specific formulas that are used in an application called variation.

OBJECTIVES

When you have finished this unit you should be able to:

1. Evaluate a formula given numerical values for the variables.
2. Solve a formula for a given variable.
3. Solve applied problems involving direct and inverse variation.

9.1 Evaluating Formulas

A **formula** is a type of equation in two or more variables. Generally, the left-hand side of a formula consists of a single variable. The right-hand side is an expression containing one or more variables. These are some examples of formulas:

$$y = 3x, \qquad D = RT, \qquad P = 2w + 2l.$$

In Section 1.2 we evaluated algebraic expressions for given values of the variables. We **evaluate a formula** by evaluating the expression on the right-hand side.

EXAMPLE 9.1 Evaluate $x = 3y - 2z$ for $y = -3$ and $z = 4$.

Solution We substitute -3 for y and 4 for z in the expression on the right-hand side of the formula:

$$x = 3y - 2z$$
$$x = 3(-3) - 2(4)$$
$$x = -9 - 8$$
$$x = -17.$$

The value of x is -17.

Letters in a formula often have special meanings. An important formula in mathematics is $y = mx + b$. In this formula, m and b have special meanings, which you will learn in Unit 10.

EXAMPLE 9.2 Evaluate the formula $y = mx + b$ for $b = -5$, $m = -\frac{1}{2}$, and $x = -10$.

Solution We substitute the values for the variables on the right-hand side, being careful to put each in its correct place:

$$y = mx + b$$
$$y = \left(-\frac{1}{2}\right)(-10) + (-5)$$
$$y = 5 - 5$$
$$y = 0.$$

The value of y is 0.

Some formulas can involve only positive values because of their nature. For example, the simple interest formula is $I = PRT$, where I is an amount of interest earned (on an investment) or owed (on a loan), P is the principal, R is the interest rate, and T is the amount of time. All of the variables must be positive for the formula to have any meaning.

EXAMPLE 9.3 Evaluate the simple interest formula if $P = \$500$, $R = 0.12$, and $T = \frac{1}{4}$.

Solution Substituting the values for the variables,

$$I = PRT$$
$$I = (500)(0.12)\left(\frac{1}{4}\right)$$
$$I = 15.$$

The interest is $15.

EXAMPLE 9.4 The formula for the circumference of a circle is $C = 2\pi r$. Evaluate the formula for $\pi = 3.1416$ and $r = 2.67$.

Solution Substituting the given values,

$$C = 2\pi r$$
$$C = 2(3.1416)(2.67)$$
$$C = 16.8$$

where we have rounded to one decimal place. You may recognize that π (the Greek letter "pi") represents a constant that is very commonly used in mathematics and science. In particular, you might have used π in formulas for the circumference and area of a circle. The constant π has many other uses.

There are also formulas from science where negative values have meaning. Two common formulas are those that convert temperatures from Fahrenheit to Celsius and from Celsius to Fahrenheit. A temperature can be negative, meaning below zero.

EXAMPLE 9.5 Evaluate $C = \frac{5}{9}(F - 32)$ for $F = -4$.

Solution   Substituting for F,

$$C = \frac{5}{9}(F - 32)$$

$$C = \frac{5}{9}(-4 - 32)$$

$$C = \frac{5}{9}(-36)$$

$$C = -20.$$

A formula may occasionally be stated in a form in which a variable is not quite isolated on the left-hand side. For example, $\frac{1}{R} = \frac{1}{R_1} + \frac{1}{R_2}$ is a formula from electronics for the combined resistance of two resistors in a parallel circuit. The symbols R, R_1, and R_2 all represent resistances. The numbers 1 and 2 to the lower right of the letters are called **subscripts.** The subscripts distinguish R_1 and R_2 from each other and from R.

EXAMPLE 9.6   Evaluate $\frac{1}{R} = \frac{1}{R_1} + \frac{1}{R_2}$ for R, if $R_1 = 8$ and $R_2 = 12$.

Solution   Substituting for R_1 and R_2,

$$\frac{1}{R} = \frac{1}{R_1} + \frac{1}{R_2}$$

$$\frac{1}{R} = \frac{1}{8} + \frac{1}{12}$$

$$\frac{1}{R} = \frac{3}{24} + \frac{2}{24}$$

$$\frac{1}{R} = \frac{5}{24}.$$

We have found $\frac{1}{R}$. To find R, we take the reciprocal of $\frac{5}{24}$. Therefore,

$$R = \frac{24}{5}$$

or

$$R = 4.8.$$

Exercise 9.1

Evaluate:

1. $z = 6x + 5y$ for $x = 3$ and $y = -2$

2. $y = -2s + 4t$ for $s = 9$ and $t = 6$

3. $p = 3r - 4q + 5s$ for $r = 10$, $q = -9$, and $s = 0$

4. $x = 6a + 4b - 2c$ for $a = -8$, $b = -4$, and $c = -3$

5. $y = mx + b$ for $m = -2$, $x = 1$, and $b = 5$

6. $y = mx + b$ for $b = -4$, $m = \frac{2}{3}$, and $x = 2$

7. $I = PRT$ for $P = \$1000$, $R = 0.08$, and $T = 5$

8. $I = PRT$ for $P = \$6000$, $R = 0.125$, and $T = \frac{1}{4}$

9. $A = P(1 + RT)$, the total amount accumulated by simple interest, for $P = \$500$, $R = 0.12$, and $T = \frac{1}{4}$

10. $A = P(1 + RT)$ for $P = \$750$, $R = 0.055$, and $T = 3$

11. $P = 2w + 2l$, the perimeter of a rectangle, for $w = 3.6$ and $l = 4.5$

12. $A = \frac{1}{2}bh$, the area of a triangle, for $b = 9$ and $h = 7$

13. $C = 2\pi r$, for $\pi = 3.1416$ and $r = 2.5$, to one decimal place

14. $C = 2\pi r$, for $\pi = 3.1416$ and $r = 4.75$, to one decimal place

15. $C = \frac{5}{9}(F - 32)$ for $F = 23$

16. $C = \frac{5}{9}(F - 32)$ for $F = -13$

17. $F = \frac{9C}{5} + 32$ for $C = -15$

18. $F = \frac{9C}{5} + 32$ for $C = -50$

19. $\frac{1}{R} = \frac{1}{R_1} + \frac{1}{R_2}$ for $R_1 = 3$ and $R_2 = 4$

20. $\frac{1}{R} = \frac{1}{R_1} + \frac{1}{R_2}$ for $R_1 = 6$ and $R_2 = 9$

9.2 Solving Formulas

To **solve a formula** for a given variable means to isolate the given variable on one side of the equal sign, with all the constants and other variables on the other side. We use the same methods as we use to isolate the variable in solving equations.

EXAMPLE 9.7   Solve $D = RT$ for T.

Solution   To isolate T, we divide both sides of the equation by R:

$$D = RT$$

$$\frac{D}{R} = \frac{RT}{R}$$

$$\frac{D}{R} = T$$

or

$$T = \frac{D}{R}.$$

EXAMPLE 9.8   Solve $P = \frac{F}{A}$, Pascal's principle for pressure in a fluid, for F.

Solution   To isolate F, we multiply both sides of the equation by A:

$$P = \frac{F}{A}$$

$$PA = \frac{FA}{A}$$

$$PA = F$$

or

$$F = PA.$$

EXAMPLE 9.9   Solve $y = mx + b$ for x.

Solution   First, we subtract b from both sides:

$$y = mx + b$$
$$y - b = mx + b - b$$
$$y - b = mx.$$

Now, we divide both sides by m:

$$\frac{y - b}{m} = \frac{mx}{m}$$

$$\frac{y - b}{m} = x$$

or

$$x = \frac{y - b}{m}.$$

There are many occasions in mathematics when an equation in the two variables x and y must be solved for y.

EXAMPLE 9.10   Solve $x + 3y - 9 = 0$ for y.

Solution   We add 9 to both sides:

$$x + 3y - 9 = 0$$
$$x + 3y = 9,$$

and subtract x from both sides:

$$3y = 9 - x.$$

Then, dividing by 3,

$$y = \frac{9 - x}{3}.$$

EXAMPLE 9.11   Solve $2x - 3y - 6 = 0$ for y.

Solution   Since the y term is negative, the easiest method of solution is to add $3y$ to both sides:

$$2x - 3y - 6 = 0$$
$$2x - 6 = 3y.$$

Then, dividing by 3,

$$\frac{2x - 6}{3} = y$$

or

$$y = \frac{2x - 6}{3}.$$

The equation may also be solved for y by subtracting $2x$ and adding 6 to both sides:

$$2x - 3y - 6 = 0$$
$$-3y = -2x + 6.$$

Now, we divide by -3:

$$y = \frac{-2x + 6}{-3}.$$

This solution is equivalent to our first solution. If we multiply both the numerator and the denominator by -1, we have

$$y = \frac{-(-2x + 6)}{3}$$
$$y = \frac{2x - 6}{3},$$

which is identical to the first solution.

Many formulas contain a factor that is an expression. If the variable for which the formula is to be solved is not in this factor, we may divide both sides by the entire factor.

EXAMPLE 9.12   Solve $V = V_0(1 + \beta t)$, expansion of a gas under constant pressure, for V_0.

Solution The subscript 0 in V_0 distinguishes V_0 from V. β (the Greek letter "beta") represents a constant. To solve for V_0, we divide both sides by the entire factor $1 + \beta t$:

$$V = V_0(1 + \beta t)$$

$$\frac{V}{1 + \beta t} = \frac{V_0(1 + \beta t)}{1 + \beta t}$$

We divide out the factor $1 + \beta t$ from the numerator and the denominator of the right-hand side. Thus,

$$\frac{V}{1 + \beta t} = V_0$$

or

$$V_0 = \frac{V}{1 + \beta t}.$$

If the variable we are solving for is in a factor in parentheses, we must remove the parentheses before attempting to solve.

EXAMPLE 9.13   Solve $V = V_0(1 + \beta t)$ for β.

Solution Since β is in the parentheses, we must first remove the parentheses:

$$V = V_0(1 + \beta t)$$
$$V = V_0 + V_0\beta t.$$

Now, we can subtract V_0 from both sides:

$$V - V_0 = V_0\beta t,$$

and divide both sides by V_0 and t:

$$\frac{V - V_0}{V_0 t} = \frac{V_0\beta t}{V_0 t}$$

$$\frac{V - V_0}{V_0 t} = \beta$$

or

$$\beta = \frac{V - V_0}{V_0 t}.$$

Exercise 9.2

Solve for the variable indicated:

1. $P = RB$ for R

2. $A = lw$, the area of a rectangle, for l

3. $I = PRT$ for R

4. $A = \frac{1}{2}bh$, the area of a triangle, for h

5. $I = \frac{E}{R}$, Ohm's law for electrical current, for E

6. $k = \frac{F}{x}$, Hooke's law for elasticity, for F

7. $I = \frac{E}{R}$ for R

8. $\mu = \frac{F}{W}$, equation for coefficient of friction (μ is the Greek letter "mu"), for W

9. $v = v_0 + at$, velocity of a falling object, for t

10. $P = 2w + 2l$ for l

Solve for y:

11. $2x + 5y - 9 = 0$

12. $6x + 3y + 4 = 0$

13. $4x - 6y - 5 = 0$

14. $3x - 4y + 7 = 0$

15. $3y - 2x - 10 = 0$

16. $8 - 6y - 5x = 0$

Solve for the variable indicated:

17. $A = P(1 + rt)$ for P

18. $A = \frac{1}{2}h(B + b)$, the area of a trapezoid, for h

19. $A = P(1 + rt)$ for t

20. $A = \frac{1}{2}h(B + b)$ for B

⌐ 9.3 ¬ Variation Formulas

Two or more variables are sometimes related according to formulas of a basic type. For example, as the speed of an automobile increases, the distance it can cover in a given time increases. This relationship is an example of **direct variation.** We define direct variation by a formula in the variables x and y, and a constant k.

Direct Variation: The relationship y varies directly as x is represented by the formula

$$y = kx$$

where k is a constant.

The constant k is called the **constant of variation.** When we are given a variation relationship, we can write the variation formula in terms of the constant of variation.

EXAMPLE 9.14   Distance D varies directly as rate of speed R. Write the formula for this variation relationship.

Solution

In the direct variation formula, the first variable is equal to the constant of variation multiplied by the second variable. Therefore, we write

$$D = kR.$$

In direct variation relationships, when the second variable increases, the first variable also increases. There is also a common type of relationship in which, when the second variable increases, the first decreases. For example, as the speed of an automobile increases, the time it takes to cover a given distance decreases. This relationship is an example of **inverse variation.**

Inverse Variation: The relationship y varies inversely as x is represented by the formula

$$y = \frac{k}{x}$$

where k is a constant.

Again, k is called the constant of variation.

EXAMPLE 9.15   Rate of speed R varies inversely as time T. Write the formula for this variation relationship.

Solution

In the inverse variation formula, the first variable is equal to the constant of variation divided by the second variable. Therefore, we write

$$R = \frac{k}{T}.$$

There are variation relationships involving more than two variables.

EXAMPLE 9.16   The volume V of a room varies directly as the length l and the width w. Write the variation formula.

Solution

Using the direct variation formula for both l and w, we multiply the constant of variation by both l and w. Thus, the formula is

$$V = klw.$$

EXAMPLE 9.17   The acceleration a of an object varies directly as the force F and inversely as the weight w. Write the variation formula.

Solution

Using the direct variation formula, we multiply the constant of variation by F. Then, using the inverse variation formula, we divide by w. The formula is

$$a = \frac{kF}{w}.$$

If we are given values for the variables in a variation relationship, we can find the value of the constant of variation. Then, we can write the variation formula in terms of the variables only.

EXAMPLE 9.18   Suppose A varies directly as h, and $A = 60$ when $h = 12$. Find k, and write the variation formula in terms of the variables only.

Solution   First, we write the variation formula in terms of k:

$$A = kh.$$

Then, we substitute the given values for A and h, so that

$$60 = k(12).$$

Solving for k,

$$\frac{60}{12} = k$$
$$5 = k$$

or

$$k = 5.$$

Therefore, the variation formula in terms of A and h is

$$A = 5h.$$

EXAMPLE 9.19   Suppose z varies inversely as x, and $z = 10$ when $x = -3$. Find k, and write the variation formula in terms of the variables only.

Solution   The variation formula is

$$z = \frac{k}{x}.$$

Substituting the values for z and x, and solving for k,

$$10 = \frac{k}{-3}$$
$$(-3)(10) = k$$
$$-30 = k$$

or

$$k = -30.$$

Therefore, the variation formula in terms of z and x is

$$z = \frac{-30}{x}.$$

EXAMPLE 9.20   The surface area S of a cylinder (a surface shaped like a tin can without the top and bottom) varies directly as the radius r and the height h. If $S = 339$ square centimeters when $r = 4.5$ centimeters and $h = 12$ centimeters, find k to two decimal places, and write the variation formula in terms of the variables only.

Solution   Since S varies directly as both r and h, the variation formula is

$$S = krh.$$

Substituting the values for S, r, and h, and solving for k,

$$339 = k(4.5)(12)$$
$$339 = k(54)$$
$$\frac{339}{54} = k$$
$$6.2777 \ldots = k$$

or

$$k = 6.28$$

to two decimal places. Therefore, the variation formula in terms of S, r, and h is

$$S = 6.28rh.$$

Finally, we can use a variation formula to find a value for one of the variables. First, to find k, we must be given a set of values for all of the variables. Then, we must be given new values for all of the variables except the one we are to find.

EXAMPLE 9.21   Suppose A varies directly as h and $A = 60$ when $h = 12$. Find A when $h = 15$.

Solution   In Example 9.18 we used the variation formula

$$A = kh$$

with $A = 60$ and $h = 12$ to find the formula

$$A = 5h.$$

Now, we substitute $h = 15$ to evaluate the formula for A:

$$A = 5(15)$$
$$A = 75.$$

EXAMPLE 9.22   Suppose y varies directly as x and inversely as t. If $y = 6$ when $x = -20$ and $t = -10$, find y when $x = -2$ and $t = 4$.

Solution   First, we write the variation formula

$$y = \frac{kx}{t}.$$

Next, using the first set of values for the variables, we find k:

$$6 = \frac{k(-20)}{-10}$$
$$6 = k(2)$$
$$3 = k,$$

and so

$$y = \frac{3x}{t}.$$

Finally, we use the second set of values for x and t to evaluate the formula:

$$y = \frac{3(-2)}{4}$$
$$y = \frac{-6}{4}$$
$$y = -\frac{3}{2}.$$

EXAMPLE 9.23   According to a law of physics, if the density of a liquid in a pipe is constant, the pressure P varies directly as the height h. If $P = 5.2$ pounds per square inch when $h = 144$ inches, find P when $h = 200$ inches.

Solution   The density is our constant k. Since P varies directly as h, we have

$$P = kh.$$

Substituting the first set of values and solving for k,

$$5.2 = k(144)$$
$$\frac{5.2}{144} = k$$
$$0.036 = k$$

(where k is a nonterminating decimal that we have rounded to three decimal places). Then, evaluating the formula for $h = 200$,

$$P = 0.036h$$
$$P = 0.036(200)$$
$$P = 7.2 \text{ pounds per square inch.}$$

Exercise 9.3

Write the variation formula:

1. Work W varies directly as distance d.

2. Circumference c varies directly as radius r.

3. Acceleration a varies inversely as mass m.

4. Current I varies inversely as resistance R.

5. The surface area S of a cylinder varies directly as the radius r and height h.

6. The gravitational force G between two objects a fixed distance apart varies directly as the mass M_1 of the first object and the mass M_2 of the second.

7. The resistance R in a wire varies directly as the length l of the wire and inversely as the cross-sectional area A.

8. The pressure P on the surface of a liquid varies directly as the force F perpendicular to the surface and inversely as the area A of the surface.

Find k, and write the variation formula in terms of the variables only:

9. D varies directly as T, and $D = 88$ when $T = 2$.

10. F varies directly as x, and $F = -6$ when $x = 12$.

11. R varies inversely as T, and $R = 30$ when $T = 2$.

12. y varies inversely as x, and $y = -10$ when $x = -5$.

13. z varies directly as x and y, and $z = -40$ when $x = -4$ and $y = -5$.

14. A varies directly as b and h, and $A = 48$ when $b = 12$ and $h = 8$.

15. p varies directly as m and inversely as n, and $p = -20$ when $m = 25$ and $n = -5$.

16. The pressure P of a gas varies directly as the absolute temperature T and inversely as the volume V, and $P = 10$ atmospheres when $T = 293°$ Kelvin and $V = 610$ cubic inches.

Solve for the indicated variable:

17. D varies directly as T, and $D = 110$ when $T = 2$. Find D when $T = 3$.

18. R varies inversely as T, and $R = 30$ when $T = 2$. Find R when $T = 5$.

19. Suppose r varies directly as s and inversely as t. If $r = 25$ when $s = -15$ and $t = 5$, find r when $s = 9$ and $t = 20$.

20. Suppose z varies directly as x and y. If $z = 24$ when $x = -3$ and $y = -6$, find z when $x = -6$ and $y = -10$.

21. If the mass of an object is constant, the force F acting on it varies directly as the acceleration a. If $F = 50$ dynes when $a = \frac{2}{30}$ centimeters per square second, find F when $a = \frac{1}{10}$ centimeters per square second.

22. If the cylindrical surface of a tin can is constant, the height h varies inversely as the diameter d. If $h = 15$ centimeters when $d = 8.5$ centimeters, find h when $d = 12$ centimeters.

23. The potential energy E of an object varies directly as the mass m and the height h. If $E = 64.68$ joules when $m = 2.2$ kilograms and $h = 3$ meters, find E when $m = 3.6$ kilograms and $h = 2$ meters.

24. The resistance R of a wire varies directly as the length l and inversely as the cross-sectional area A. If $R = 0.005$ ohms when $l = 100$ feet and $A = 0.0314$ square inches, find R when $l = 200$ feet and $A = 0.00785$ square inches.

Self-test

Evaluate:

1. $y = 2a - 3b - c$ for $a = 6$, $b = 0$, and $c = -5$

1. _____

2. $F = \dfrac{9C}{5} + 32$ for $C = -10$

2. _____

Solve for the variable indicated:

3. $A = P(1 + rt)$ for r

3. _____

4. $5x + 6y - 12 = 0$ for y

4. _____

5. x varies directly as y and inversely as z. If $x = 10$ when $y = -4$ and $z = 2$, find x when $y = 15$ and $z = 25$.

5. _____

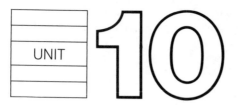

UNIT 10 — Linear Equations in Two Variables

INTRODUCTION

In earlier units you have used equations with one variable. In the preceding unit you learned about formulas, which are a type of equation with two or more variables. Such equations have many solutions, so many that it is not possible to write them all down. Renè Descartes invented a way to indicate all the solutions of an equation in two variables. The method is called the Cartesian coordinate system in his honor. In this unit you will learn how to use the Cartesian coordinate system, how to draw the graph of a line, and some properties of graphs of lines.

OBJECTIVES

When you have finished this unit you should be able to:

1. Plot a point with given coordinates in the Cartesian coordinate system.
2. Draw the graph of a line, given the equation.
3. Find the y- and x-intercepts of a line and use them to draw the graph.
4. Find the slope of a line determined by two given points.
5. Find the slope and y-intercept of a line, given its equation.

10.1 The Cartesian Coordinate System

To construct the **Cartesian coordinate system** we use the number line introduced in Section 3.1. Suppose we have two number lines placed perpendicularly; that is, at right angles to one another. It is usual to place one number line horizontally and the other vertically:

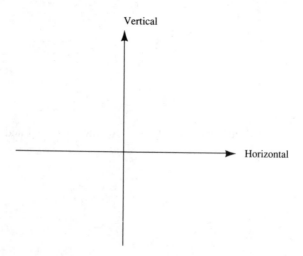

It is also usual to direct the horizontal number line to the right and the vertical number line upward. The origin of each number line is taken to be the point where the lines cross.

The common origin of the two number lines is called the **origin** of the Cartesian coordinate system. The number lines themselves are called **axes.** In this unit, we will call the horizontal number

134

line the **x-axis** and the vertical number line the **y-axis.** We label the axes x and y and rule off coordinates on each:

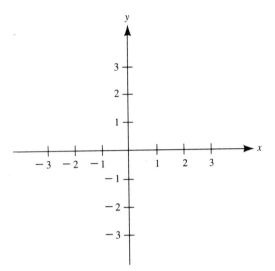

In specific situations in mathematics and science variables other than x and y are used.

In Section 3.1 we drew the graph of a number as a point on the number line. We call the number the coordinate of the point. In the Cartesian coordinate system the location of a point is given by two numbers. We always write the two numbers as an **ordered pair,** such as (2, 3). In an ordered pair, the numbers are written in parentheses and separated by a comma. The numbers in an ordered pair are called **coordinates.** The first number is called the **x-coordinate** and the second number is called the **y-coordinate.** In the ordered pair (2, 3), the x-coordinate is 2 and the y-coordinate is 3.

The word "ordered" is important because the order of the coordinates is important. The ordered pair (3, 2), like the pair (2, 3), has coordinates 2 and 3. However, in the ordered pair (3, 2) the x-coordinate is 3 and the y-coordinate is 2.

Traditionally, the x and y coordinates were called the **abscissa** and the **ordinate.** These two names are no longer encountered frequently. We will simply use the variable names x and y, or x-coordinate and y-coordinate.

To graph an ordered pair in the Cartesian coordinate system, we locate the x-coordinate on the x-axis and the y-coordinate on the y-axis. Then, starting from the x-coordinate, we go up or down until we are level with the y-coordinate. The graph of (2, 3) is

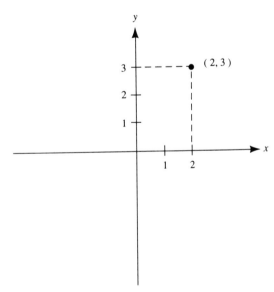

The graph of (3, 2) is

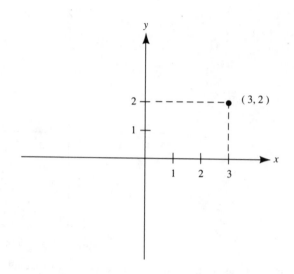

Observe that the graph of (2, 3) is not the same as the graph of (3, 2).

The dashed lines in the previous diagrams are only guidelines, and we will not generally show them. Observe, however, that the dashed lines, together with the axes, form a rectangle. For this reason the Cartesian coordinate system is also commonly called the **rectangular coordinate system.**

EXAMPLE 10.1   Graph the point (6, 4) in the Cartesian coordinate system.

Solution   We locate 6 on the *x*-axis and 4 on the *y*-axis. Then, starting from 6 we go up until we are level with 4:

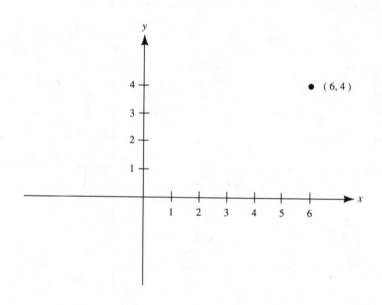

If the x-coordinate is negative, we go to the left of the origin. If the y-coordinate is negative, we go down from the origin.

EXAMPLE 10.2   Graph and label each point in the Cartesian coordinate system:

a. $(-3, 5)$ b. $(5, -2)$ c. $(-2, -4)$

Solutions   a. We go left from the origin to -3 on the x-axis, then up until we are level with 5 on the y-axis.
b. We go down from 5 on the x-axis until we are level with -2 on the y-axis.
c. We go left on the x-axis to -2, then down until we are level with -4 on the y-axis.

The graph is

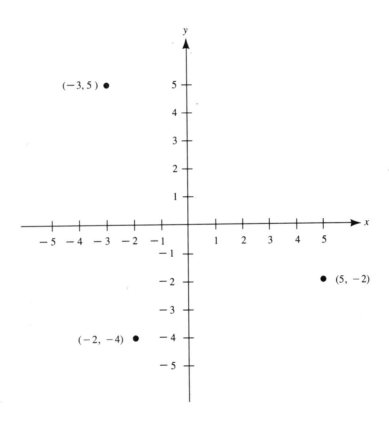

The graph of $(0, 0)$ is the origin. If one coordinate of a point is 0, the graph of the point lies on one of the axes.

EXAMPLE 10.3   Graph and label each point in the Cartesian coordinate system:

a. $(6, 0)$ b. $(0, -5)$

Solutions   a. We locate 6 on the x-axis. Since the y-coordinate is 0, we do not go up or down at all. The point is on the x-axis.
b. Since the x-coordinate is 0, we do not go left or right at all. The point is at -5 on the y-axis.

The graph is

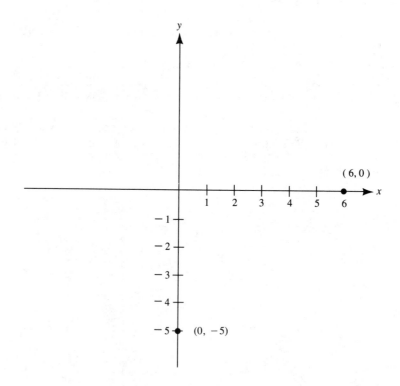

It is a common error to graph an ordered pair such as $(-8, 7)$ as two points, -8 on the x-axis and 7 on the y-axis. However, $(-8, 7)$ represents just one point, where the x-coordinate is -8 and the y-coordinate is 7. The point at -8 on the x-axis is $(-8, 0)$ and the point at 7 on the y-axis is $(0, 7)$.

EXAMPLE 10.4   Graph and label each point in the Cartesian coordinate system:

a. $(-8, 7)$ b. $(-8, 0)$ c. $(0, 7)$

Solutions   Proceeding as in the previous examples, construct the graph on page 139.

Exercise 10.1

Graph and label each point in the Cartesian coordinate system:

1. a. $(3, 4)$ b. $(2, -5)$ c. $(-1, 6)$ d. $(-3, -6)$

2. a. $(5, 2)$ b. $(-4, 1)$ c. $(4, -5)$ d. $(-5, -3)$

3. a. $(1, 0)$ b. $(0, 2)$ c. $(-3, 0)$ d. $(0, -5)$

4. a. $(5, 0)$ b. $(-6, 0)$ c. $(0, 4)$ d. $(0, -2)$

5. a. $(-4, -2)$ b. $(-4, 0)$ c. $(0, -2)$

6. a. $(8, -10)$ b. $(8, 0)$ c. $(0, -10)$

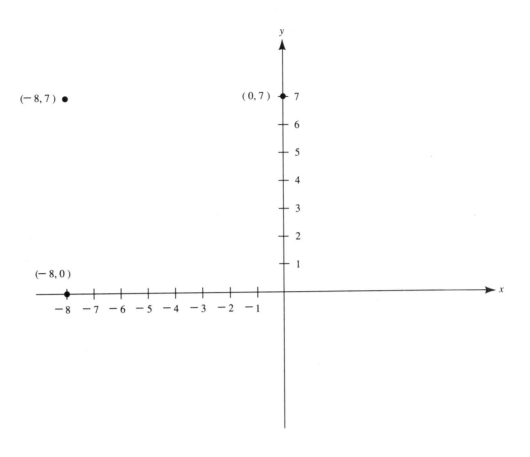

10.2 Graphs of Linear Equations

In Unit 2 we gave some examples of linear equations in one variable. These are examples of linear equations in two variables:

$$y = \frac{1}{2}x + 2, \quad 3x + 2y - 6 = 0.$$

A **solution** of an equation in two variables x and y is an ordered pair that makes the two sides of the equation numerically equal when the x-coordinate of the ordered pair is substituted for x and the y-coordinate is substituted for y. Consider, for example, the equation $y = \frac{1}{2}x + 2$. The ordered pair (2, 3) is a solution of this equation because, when 2 is substituted for x and 3 is substituted for y, the result is

$$y = \frac{1}{2}x + 2$$

$$3 \stackrel{?}{=} \frac{1}{2}(2) + 2$$

$$3 \stackrel{?}{=} 1 + 2$$

$$3 = 3.$$

However, the ordered pair (3, 2) is not a solution of the equation because

$$y = \frac{1}{2}x + 2$$

$$2 \stackrel{?}{=} \frac{1}{2}(3) + 2$$

$$2 \stackrel{?}{=} \frac{3}{2} + 2$$

$$2 \neq \frac{7}{2}.$$

The equation $y = \frac{1}{2}x + 2$ has many other solutions besides (2, 3). You should check each of these sample solutions:

$$(0, 2), \quad \left(1, \frac{5}{2}\right), \quad \left(3, \frac{7}{2}\right), \quad \left(-1, \frac{3}{2}\right), \quad (-2, 1).$$

We could continue a list of solutions indefinitely. We say that the equation has **infinitely many** solutions. These solutions do not just include ordered pairs where x is an integer; there is an ordered pair solution for any real number x.

Clearly, we cannot list all of the solutions of a linear equation in two variables. We indicate the solutions by drawing the **graph of the equation.** To draw the graph of the equation $y = \frac{1}{2}x + 2$, we first draw the graphs of the solutions we have found:

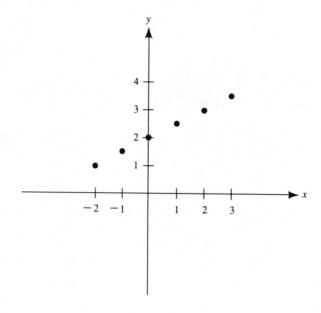

The solutions appear to form a straight line. If we fill in the line, its points represent all the solutions of the equation:

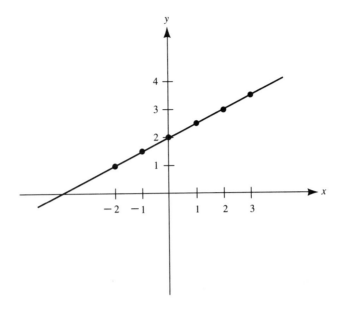

The line continues indefinitely in both directions. If we continue our list of solutions, we will find where the line crosses the x-axis:

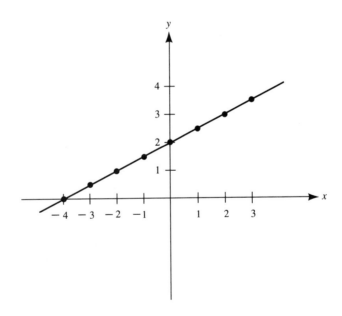

Whenever possible, we will continue lines until they cross both axes.

The graph of a linear equation in two variables is always a straight line. The name "linear" derives from this fact.

EXAMPLE 10.5   Draw the graph of $y = 2x - 1$.

Solution   We must find some ordered pairs that are solutions of the equation. In this section, we will generally start with $x = 0$. If $x = 0$,

$$
\begin{aligned}
y &= 2x - 1 \\
&= 2(0) - 1 \\
&= 0 - 1 \\
&= -1.
\end{aligned}
$$

Therefore, one solution is $(0, -1)$. Now, we might go to $x = 1$:

$$
\begin{aligned}
y &= 2x - 1 \\
&= 2(1) - 1 \\
&= 2 - 1 \\
&= 1.
\end{aligned}
$$

Another solution is $(1, 1)$. Two points are sufficient to determine a straight line; however, we will find a third point as a check point. If $x = 2$,

$$
\begin{aligned}
y &= 2x - 1 \\
&= 2(2) - 1 \\
&= 4 - 1 \\
&= 3.
\end{aligned}
$$

A third solution is $(2, 3)$. You may want to keep track of the solutions on a chart like this:

x	y
0	-1
1	1
2	3

We graph our three points and draw a straight line through them:

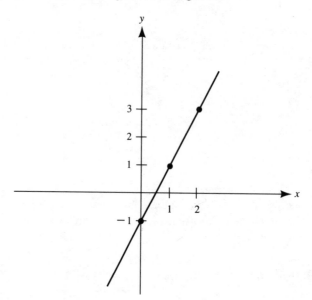

| | EXAMPLE 10.6   Draw the graph of $y = -x - 3$.

Solution   If $x = 0$,

$$y = -x - 3$$
$$= -0 - 3$$
$$= -3.$$

If $x = 1$,

$$y = -x - 3$$
$$= -1 - 3$$
$$= -4.$$

If $x = 2$,

$$y = -x - 3$$
$$= -2 - 3$$
$$= -5.$$

x	y
0	-3
1	-4
2	-5

Three solutions are $(0, -3)$, $(1, -4)$ and $(2, -5)$:

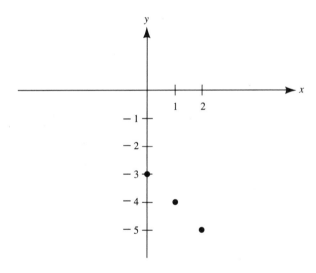

Observe that these solutions are dropping away from the x-axis. We should try some negative values for x. If $x = -1$,

$$y = -x - 3$$
$$= -(-1) - 3$$
$$= 1 - 3$$
$$= -2.$$

If $x = -2$,

$$y = -x - 3$$
$$= -(-2) - 3$$
$$= 2 - 3$$
$$= -1.$$

If $x = -3$,

$$y = -x - 3$$
$$= -(-3) - 3$$
$$= 3 - 3$$
$$= 0.$$

x	y
0	-3
1	-4
2	-5
-1	-2
-2	-1
-3	0

Three more solutions are $(-1, -2)$, $(-2, -1)$, and $(-3, 0)$. The graph is

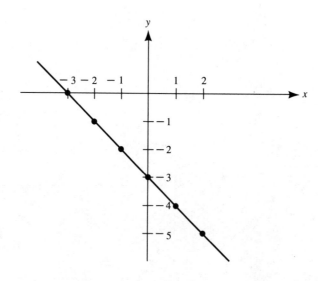

Usually, it is not necessary to find so many solutions. Three representative solutions such as $(0, -3)$, $(1, -4)$ and $(-3, 0)$ would be sufficient.

If a linear equation in two variables is not given with y isolated, it is often convenient to solve the equation for y.

EXAMPLE 10.7   Draw the graph of $x + y = 1$.

Solution   First, we solve the equation for y:

$$x + y = 1$$
$$y = 1 - x.$$

Then, we find three representative solutions. If $x = 0$,

$$y = 1 - x$$
$$= 1 - 0$$
$$= 1.$$

If $x = 1$,

$$y = 1 - x$$
$$= 1 - 1$$
$$= 0.$$

If $x = -1$,

$$y = 1 - x$$
$$= 1 - (-1)$$
$$= 1 + 1$$
$$= 2.$$

x	y
0	1
1	0
-1	2

Three representative solutions are $(0, 1)$, $(1, 0)$, and $(-1, 2)$. The graph is

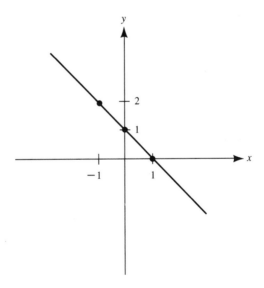

The choice of $x = -1$ was arbitrary. We could just as well have chosen, for example, $x = 2$. You should find this solution and graph it.

EXAMPLE 10.8 Draw the graph of $3x + 2y - 6 = 0$.

Solution   Solving the equation for y,

$$3x + 2y - 6 = 0$$
$$2y = 6 - 3x$$
$$y = \frac{6 - 3x}{2}.$$

It is often convenient to write the right-hand side with separate denominators:

$$y = \frac{6}{2} - \frac{3x}{2}$$

$$y = 3 - \frac{3}{2}x.$$

To make it easier to calculate solutions, we choose values of x that will give integers. If $x = 0$,

$$y = 3 - \frac{3}{2}x$$

$$= 3 - \frac{3}{2}(0)$$

$$= 3.$$

If $x = 2$,

$$y = 3 - \frac{3}{2}(2)$$

$$= 3 - 3$$

$$= 0.$$

If $x = -2$,

$$y = 3 - \frac{3}{2}(-2)$$

$$= 3 + 3$$

$$= 6.$$

x	y
0	3
2	0
-2	6

Three solutions are $(0, 3)$, $(2, 0)$, and $(-2, 6)$. The graph is

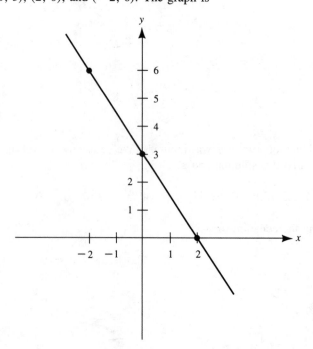

There are two special cases of linear equations in two variables. These are the cases in which one of the variables does not appear.

EXAMPLE 10.9   Draw the graph of $y = 3$.

Solution   The variable x does not appear. However, x can have any real-number value. According to the equation, $y = 3$ for any value of x. Some solutions are $(0, 3)$, $(1, 3)$, and $(-1, 3)$. The graph is

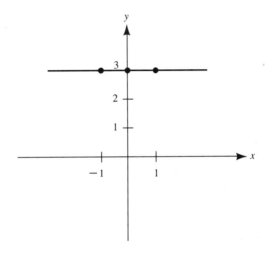

Whenever x does not appear in a linear equation, the equation can be written in the form $y = a$, and the graph of the equation is a horizontal line.

EXAMPLE 10.10   Draw the graph of $2x + 3 = 0$.

Solution   Since y does not appear, we solve the equation for x:

$$2x + 3 = 0$$
$$2x = -3$$
$$x = -\frac{3}{2}.$$

The variable x can have only one value, $x = -\frac{3}{2}$. However, y can have any real-number value. Some solutions are $\left(-\frac{3}{2}, 0\right)$, $\left(-\frac{3}{2}, 1\right)$, and $\left(-\frac{3}{2}, -1\right)$. The graph is

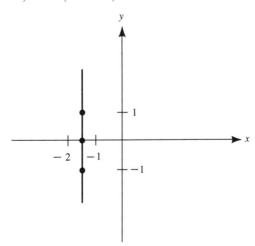

Whenever y does not appear in a linear equation, the equation can be written in the form $x = a$, and the graph of the equation is a vertical line.

Exercise 10.2

Draw the graph:

1. $y = x - 1$
2. $y = x + 3$
3. $y = 2x - 2$
4. $y = -2x + 1$

5. $y = \frac{1}{3}x + 1$
6. $y = \frac{1}{2}x - 3$
7. $y = 3x$
8. $y = -\frac{1}{2}x$

9. $x + y = 3$
10. $x - y = 2$
11. $3x - 2y + 6 = 0$
12. $x + 2y - 2 = 0$

13. $y = -2$
14. $2y - 5 = 0$
15. $x = 4$
16. $3x + 4 = 0$

10.3   Intercepts

In the preceding section, unless a line was horizontal or vertical, we continued graphs of lines until they crossed both axes. The points where a graph crosses the axes are called **intercepts.** The point where a graph crosses the y-axis is called the **y-intercept.** At the y-intercept, $x = 0$. The point where a graph crosses the x-axis is called the **x-intercept.** At the x-intercept, $y = 0$. Often, an efficient way to draw the graph of a line is by finding its intercepts.

EXAMPLE 10.11   Find the y-intercept and x-intercept, and draw the graph of $x + y = 1$.

Solution   At the y-intercept, $x = 0$. When $x = 0$,

$$x + y = 1$$
$$0 + y = 1$$
$$y = 1.$$

The y-intercept is $(0, 1)$. At the x-intercept, $y = 0$. When $y = 0$,

$$x + y = 1$$
$$x + 0 = 1$$
$$x = 1.$$

The x-intercept is $(1, 0)$. We graph the intercepts and draw the graph of the line atop page 149. This graph is the same as the one in Example 10.7.

EXAMPLE 10.12   Find the y-intercept and x-intercept, and draw the graph of $3x + 2y - 6 = 0$.

Solution   When $x = 0$,

$$3x + 2y - 6 = 0$$
$$3(0) + 2y - 6 = 0$$
$$2y - 6 = 0$$
$$2y = 6$$
$$y = 3.$$

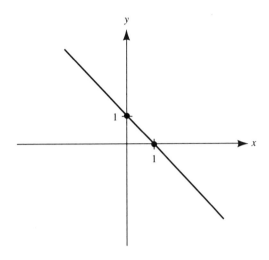

Graph for Example 10.11

The y-intercept is $(0, 3)$. When $y = 0$,

$$3x + 2y - 6 = 0$$
$$3x + 2(0) - 6 = 0$$
$$3x - 6 = 0$$
$$3x = 6$$
$$x = 2.$$

The x-intercept is $(2, 0)$. The graph is

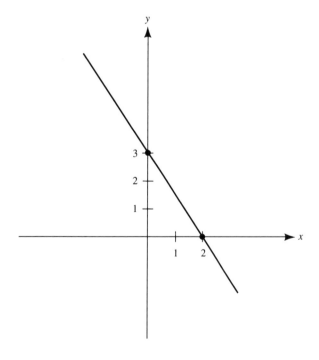

This graph is the same as the one in Example 10.8.

The two preceding graphs were checked against previous examples. In using the two intercepts to draw the graph of a line, we should find some third point as a check point.

EXAMPLE 10.13   Find the y-intercept and x-intercept, and draw the graph of $5x - 2y + 5 = 0$.

Solution   When $x = 0$,

$$5x - 2y + 5 = 0$$
$$5(0) - 2y + 5 = 0$$
$$-2y + 5 = 0$$
$$5 = 2y$$
$$\frac{5}{2} = y.$$

The y-intercept is $\left(0, \frac{5}{2}\right)$. When $y = 0$,

$$5x - 2y + 5 = 0$$
$$5x - 2(0) + 5 = 0$$
$$5x + 5 = 0$$
$$5x = -5$$
$$x = -1.$$

The x-intercept is $(-1, 0)$. To find a check point, we may choose any value other than 0 for x. If $x = 1$,

$$5x - 2y + 5 = 0$$
$$5(1) - 2y + 5 = 0$$
$$5 - 2y + 5 = 0$$
$$-2y + 10 = 0$$
$$10 = 2y$$
$$5 = y.$$

Another solution is $(1, 5)$. The graph is on the left at the top of page 151.

It is possible for the two intercepts to be identical. This happens when both intercepts are at the origin.

EXAMPLE 10.14   Find the y-intercept and x-intercept, and draw the graph of $x + 4y = 0$.

Solution   When $x = 0$,

$$x + 4y = 0$$
$$0 + 4y = 0$$
$$4y = 0$$
$$y = 0.$$

The y-intercept is $(0, 0)$. When $y = 0$,

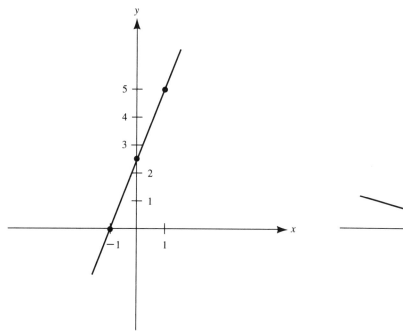

Graph for Example 10.13   Graph for Example 10.14

$$x + 4y = 0$$
$$x + 4(0) = 0$$
$$x = 0,$$

so the x-intercept also is $(0, 0)$. The graph of $(0, 0)$ is the origin. To find another solution, we choose any value for x. If $x = 4$,

$$x + 4y = 0$$
$$4 + 4y = 0$$
$$4y = -4$$
$$y = -1.$$

Therefore, another solution is $(4, -1)$. The graph is on the right at the top of this page.

We chose $x = 4$ rather than $x = 1$ in this example because $x = 4$ gives an easier solution that does not involve fractions, and because $x = 1$ would give a solution very close to the common intercept $(0, 0)$.

For equations in which either x or y does not appear, only one intercept can be determined; the other intercept does not exist.

EXAMPLE 10.15 Find the y-intercept and x-intercept, and draw the graph of $y + 5 = 0$.

Solution   Since

$$y + 5 = 0$$
$$y = -5,$$

y always has the value -5. In particular, when $x = 0$, $y = -5$. Thus the y-intercept is $(0, -5)$. But y cannot be zero, and thus there is no x-intercept. The graph is

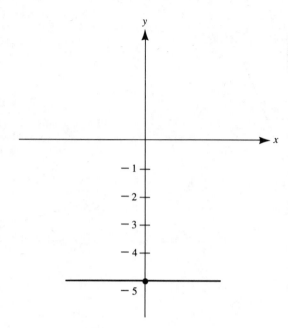

Observe that this line never crosses the x-axis.

EXAMPLE 10.16   Find the y-intercept and x-intercept, and draw the graph of $2x - 3 = 0$.

Solution x cannot be zero, so there is no y-intercept. However, for any value of y,

$$2x - 3 = 0$$
$$2x = 3$$
$$x = \frac{3}{2}.$$

Thus the x-intercept is $\left(\frac{3}{2}, 0\right)$. The graph is

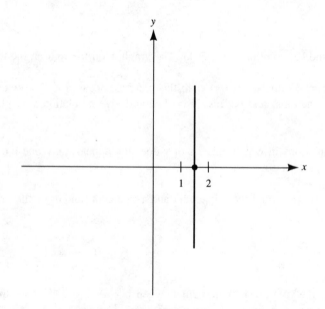

Observe that this line never crosses the y-axis.

Exercise 10.3

Find the *y*-intercept and *x*-intercept, and draw the graph:

1. $x + y = 5$ 2. $x - y = 4$

3. $x - 2y = 3$ 4. $-x + 6y = 6$

5. $3x + 4y - 12 = 0$ 6. $3x + y - 6 = 0$

7. $3x - 2y + 9 = 0$ 8. $x - 2y + 4 = 0$

9. $-2x + 5y + 10 = 0$ 10. $-7x + 2y - 14 = 0$

11. $3x + 2y = 0$ 12. $2x - 5y = 0$

13. $y - 3 = 0$ 14. $2y + 7 = 0$

15. $3x - 4 = 0$ 16. $x + 2 = 0$

10.4   Slope

In graphing lines in the two preceding units, you may have noticed that some lines rise more steeply than others:

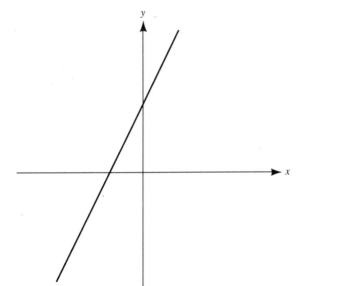
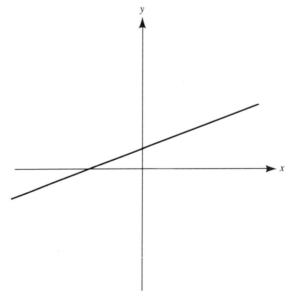

or fall more steeply than others:

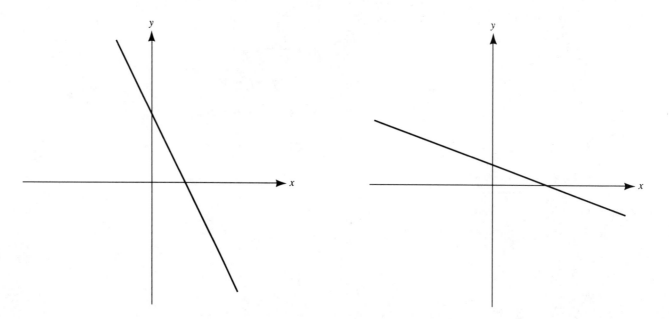

The measure of how steeply a line rises or falls is called the **slope** of the line. The slope of a line is the amount of change in the y-direction divided by the amount of change in the x-direction. We sometimes refer to this concept as "rise over run":

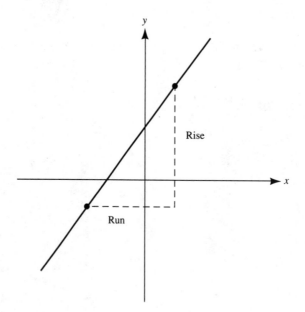

Suppose (x_1, y_1) and (x_2, y_2) are two points on a line. We use the subscripts 1 and 2 to distinguish between the x-coordinates of the two points and between the y-coordinates of the two points. We have drawn the guidelines to show the locations of the coordinates on the axes. The change in y is given by $y_2 - y_1$, and the change in x is given by $x_2 - x_1$:

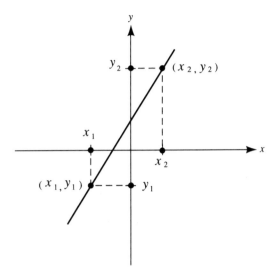

Slope Formula: The slope m of a line is given by

$$m = \frac{y_2 - y_1}{x_2 - x_1}.$$

The letter m is commonly used to represent the slope of a line. Observe that we do not use absolute values for the changes in x and y, as we did for distances in Section 3.3. A negative change will have an important interpretation.

EXAMPLE 10.17   Find the slope of the line determined by (1, 3) and (2, 5).

Solution   If we use (1, 3) for (x_1, y_1) and (2, 5) for (x_2, y_2), we have

$$m = \frac{y_2 - y_1}{x_2 - x_1}$$
$$= \frac{5 - 3}{2 - 1}$$
$$= \frac{2}{1}$$
$$= 2.$$

If we draw the line, we observe that it rises 2 units vertically for one unit of horizontal run:

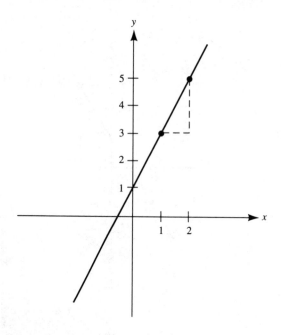

We will get the same result if we use the points in the opposite order: If we use $(2, 5)$ for (x_1, y_1) and $(1, 3)$ for (x_2, y_2), we have

$$m = \frac{y_2 - y_1}{x_2 - x_1}$$
$$= \frac{3 - 5}{1 - 2}$$
$$= \frac{-2}{-1}$$
$$= 2.$$

EXAMPLE 10.18   Find the slope of the line determined by $(2, 7)$ and $(4, 3)$.

Solution   Using $(2, 7)$ for (x_1, y_1) and $(4, 3)$ for (x_2, y_2),

$$m = \frac{y_2 - y_1}{x_2 - x_1}$$
$$= \frac{3 - 7}{4 - 2}$$
$$= \frac{-4}{2}$$
$$= -2.$$

If we draw the line, we observe that it drops 4 units vertically for every 2 units of horizontal run, or equivalently, drops 2 units for one unit run. The graph is at the top of page 157. You should check to see that you get the same result if you use the points in the opposite order in the scope formula.

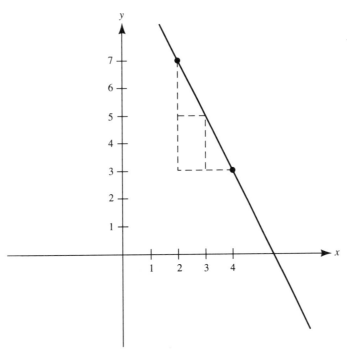

Graph for Example 10.18

We can now interpret a negative change in the slope formula. When the slope is positive, a line rises from left to right. However, when the slope is negative, a line drops from left to right.

EXAMPLE 10.19 Find the slope of the line determined by $(-1, 3)$ and $(2, 5)$.

Solution

$$m = \frac{y_2 - y_1}{x_2 - x_1}$$

$$= \frac{5 - 3}{2 - (-1)}$$

$$= \frac{2}{3}.$$

The line rises from left to right:

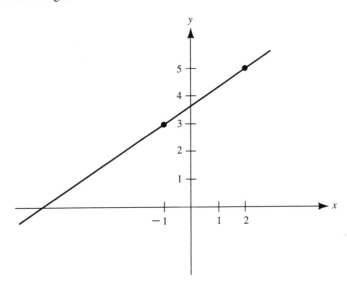

EXAMPLE 10.20   Find the slope of the line determined by $(-2, 3)$ and $(4, -1)$.

Solution

$$m = \frac{y_2 - y_1}{x_2 - x_1}$$

$$= \frac{-1 - 3}{4 - (-2)}$$

$$= \frac{-4}{6}$$

$$= -\frac{2}{3}.$$

The line drops from left to right:

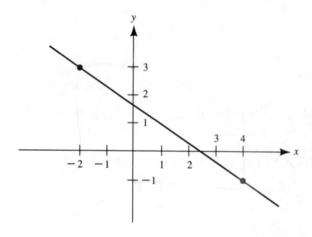

As in the preceding sections, we have two special cases.

EXAMPLE 10.21   Find the slope of the line determined by $(-3, 4)$ and $(5, 4)$.

Solution   The slope is

$$m = \frac{y_2 - y_1}{x_2 - x_1}$$

$$= \frac{4 - 4}{5 - (-3)}$$

$$= \frac{0}{8}$$

$$= 0.$$

The graph of the line is at the top of page 159. Observe that this line is horizontal. A horizontal line neither rises nor falls; its slope is zero.

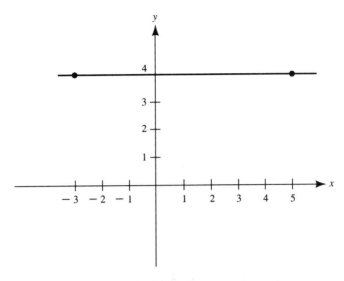

Graph for Example 10.21

EXAMPLE 10.22   Find the slope of the line determined by $(-2, 6)$ and $(-2, -1)$.

Solution   The slope is

$$m = \frac{y_2 - y_1}{x_2 - x_1}$$

$$= \frac{-1 - 6}{-2 - (-2)}$$

$$= \frac{-7}{-2 + 2}$$

$$= \frac{-7}{0}.$$

Since we cannot divide by zero, we say the slope is *undefined*. The graph of the line is

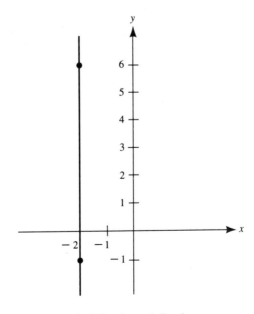

This line is vertical. The slope of a vertical line is undefined.

☐ Exercise 10.4

Find the slope of the line determined by the pair of points:

1. (1, 3) and (2, 6)   2. (6, 5) and (4, 1)   3. (2, 8) and (4, 4)   4. (6, 3) and (4, 6)

5. (−1, 5) and (4, −2)   6. (−5, 2) and (3, −8)   7. (−6, 0) and (2, −8)   8. (5, 8) and (0, −2)

9. (7, −3) and (−5, 3)   10. (2, −1) and (−3, −2) 11. (0, −6) and (2, 0)   12. (4, 0) and (0, 6)

13. (5, 3) and (5, −4)   14. (4, −2) and (6, −2)   15. (−3, −3) and (2, −3) 16. (−1, 0) and (−1, 5)

☐ 10.5 ☐ # The Slope-Intercept Form

Suppose we have a line for which the slope is defined; that is, the line is not vertical. Then, the variable y appears in the equation of the line, and we can solve the equation for y. For example, consider the line with the equation

$$y = \frac{1}{2}x + 2.$$

This equation is already solved for y. When $x = 0$,

$$y = \frac{1}{2}x + 2$$

$$= \frac{1}{2}(0) + 2$$

$$= 2.$$

The y-intercept is (0, 2). Instead of finding the x-intercept, we will find the solution for $x = 2$:

$$y = \frac{1}{2}x + 2$$

$$= \frac{1}{2}(2) + 2$$

$$= 1 + 2$$

$$= 3.$$

Thus, one solution is (2, 3). Using the points (0, 2) and (2, 3), we find the slope:

$$m = \frac{y_2 - y_1}{x_2 - x_1}$$

$$= \frac{3 - 2}{2 - 0}$$

$$= \frac{1}{2}.$$

The slope is $\frac{1}{2}$. Observe that this result is the same as the numerical coefficient of x. In general, whenever we have an equation of a line in the form

$$y = mx + b,$$

the slope is m and the y-intercept is $(0, b)$. This form is called the **slope-intercept** form of the equation.

EXAMPLE 10.23   Find the slope and the y-intercept of the line given by $2x + y - 5 = 0$.

Solution   We solve the equation for y:

$$2x + y - 5 = 0$$
$$y = -2x + 5.$$

Therefore, $m = -2$ and $b = 5$. The slope is -2, and the y-intercept is $(0, 5)$.

EXAMPLE 10.24   Find the slope and the y-intercept of the line given by $4x - 2y - 5 = 0$.

Solution   Solving the equation for y,

$$4x - 2y - 5 = 0$$
$$4x - 5 = 2y$$
$$\frac{4x - 5}{2} = y$$

or

$$y = \frac{4x - 5}{2}.$$

We write the two terms on the right-hand side with separate denominators:

$$y = \frac{4x}{2} - \frac{5}{2}$$
$$y = 2x - \frac{5}{2}.$$

Therefore, $m = 2$ and $b = -\frac{5}{2}$. The slope is 2, and the y-intercept is $\left(0, -\frac{5}{2}\right)$.

EXAMPLE 10.25   Find the slope and the y-intercept of the line given by $2x + 3y - 4 = 0$.

Solution   Solving the equation for y,

$$2x + 3y - 4 = 0$$
$$3y = -2x + 4$$
$$y = \frac{-2x}{3} + \frac{4}{3}$$
$$y = -\frac{2}{3}x + \frac{4}{3}.$$

Therefore, $m = -\frac{2}{3}$ and $b = \frac{4}{3}$. The slope is $-\frac{2}{3}$, and the y-intercept is $\left(0, \frac{4}{3}\right)$.

EXAMPLE 10.26   Find the slope and the y-intercept of the line given by $3y + 5 = 0$.

Solution   Solving the equation for y,

$$3y + 5 = 0$$
$$3y = -5$$
$$y = -\frac{5}{3}.$$

The variable x does not appear. This means the line is horizontal, and the slope is 0. The y-intercept is $\left(0, -\frac{5}{3}\right)$.

The equation of a vertical line cannot be written in slope-intercept form, because the variable y does not appear and the slope is undefined.

Exercise 10.5

Find the slope and the y-intercept of the line given by the equation:

1. $4x + y + 3 = 0$ 2. $5x - y - 6 = 0$ 3. $6x - 3y - 4 = 0$ 4. $2x - 3y - 6 = 0$

5. $2x + 4y - 5 = 0$ 6. $4x + 2y + 5 = 0$ 7. $-6x - 3y + 2 = 0$ 8. $-5x + 4y + 4 = 0$

9. $2y + 1 = 0$ 10. $6y - 5 = 0$

Self-test

1. Graph and label each point in the Cartesian coordinate system:

   a. $(4, -3)$

   b. $(-1, 5)$

   c. $(0, -6)$

   d. $(3, 0)$

2. Draw the graph of $y = 4x - 5$.

3. Find the y-intercept and x-intercept, and draw the graph of $2x - 3y + 18 = 0$.

   y-intercept _____

   x-intercept _____

4. Find the slope of the line determined by $(6, -3)$ and $(-2, -7)$.

   slope _____

5. Find the slope and the y-intercept of the line given by $3x + 4y + 5 = 0$.

   slope _____

   y-intercept _____

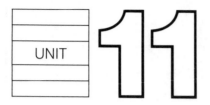

UNIT **11**

Cumulative Review

INTRODUCTION

In this unit you should make certain you remember all the material in all of the preceding units.

OBJECTIVE

When you have finished this unit you should be able to demonstrate that you can fulfill every objective of each preceding unit.

To prepare for this unit you should review the Self-Tests for Units 1 through 10. Do each problem of each Self-Test over again. If you cannot do a problem, or even if you have the slightest difficulty, you should:

1. Find out from the answer section the Objective for the unit to which the problem relates.
2. Review all the material in the section which has the same number as the objective, and redo all the Exercises for the section.
3. Try the Self-Test for the unit again.

Repeat these steps until you can do each problem in each Self-Test for each of Units 1 through 10 easily and accurately.

Self-test

1. Evaluate the expression

   $xy + xyz$

   if $x = 3$, $y = \frac{1}{3}$, and $z = \frac{2}{3}$.

   1. _____

2. Find the value of

   $10 - 2[-6 + (-9)]$

   2. _____

3. Simplify:

   $4(3t - 2) - (t - 5)$

   3. _____

Solve and check:

4. $\dfrac{3u - 4}{4} + 1 = 3$

   4. _____

5. $6 - (z - 3) = -2(z + 3)$

   5. _____

6. $-5(t - 2) = 6 - (5t - 4)$

6. _____

7. The ratio of the length of the side of any square to the length of its diagonal is approximately $1:1.414$. How long is the diagonal of a square if its side is 5.5 inches, to one decimal place?

7. _____

8. A case of canned corn and canned peas contains 24 cans. Each can of corn is worth 45¢ and each can of peas is worth 51¢. The entire case has a total value of $11.70. How many cans of corn and how many cans of peas does it contain?

8. _____

9. Solve $s = a(u - t)$ for t.

9. _____

10. Find the y-intercept and x-intercept, and draw the graph of $2x - 3y = 9$.

y-intercept _____

x-intercept _____

12 Systems of Linear Equations

INTRODUCTION

In Unit 10, you learned that a linear equation in two variables has infinitely many solutions. However, two such equations, taken together, generally have just one solution in common. In this unit you will learn how to find the solution, if any, of a system of two linear equations in two variables, and how to identify two special cases that have no solution or infinitely many solutions. Then you will learn some applications of systems of linear equations in two variables.

OBJECTIVES

When you have finished this unit you should be able to:

1. Solve systems of two linear equations in two variables by graphing.
2. Solve systems of two linear equations in two variables algebraically by the addition method.
3. Solve systems of two linear equations in two variables algebraically by the substitution method.
4. Solve applied problems using systems of two linear equations in two variables.

12.1 Solutions by Graphing

Suppose we have two lines in the Cartesian coordinate system. The two lines may intersect, may be parallel, or may be identical. The second two cases are special cases, which we will discuss at the end of this section. If the two lines intersect, then they cross at exactly one point. We call this point the **point of intersection** of the two lines.

The equations of two lines in the Cartesian coordinate system form a **system of two linear equations in two variables.** The ordered pair giving the coordinates of the point of intersection of the two lines is the **solution** of the system of equations. The solution of the system of equations is a solution of each of the individual equations. For example, consider the system of equations

$$x - y = 4$$
$$2x + 3y = 3,$$

and the ordered pair $(3, -1)$. If we substitute the coordinates of the ordered pair in the first equation, we have

$$x - y = 4$$
$$3 - (-1) \overset{?}{=} 4$$
$$3 + 1 \overset{?}{=} 4$$
$$4 = 4.$$

Therefore, the ordered pair is a solution of the first equation. If we substitute in the second equation, we have

$$2x + 3y = 3$$
$$2(3) + 3(-1) \stackrel{?}{=} 3$$
$$6 - 3 \stackrel{?}{=} 3$$
$$3 = 3.$$

Therefore, the ordered pair is a solution of the second equation. Thus, the ordered pair $(3, -1)$ is the solution of the system of equations.

To draw the **graph of the system,** we draw the graphs of both lines on one set of axes. We may use the intercepts of each line to draw its graph as in Section 10.3. For the first line of the system above, if $x = 0$,

$$x - y = 4$$
$$0 - y = 4$$
$$-y = 4$$
$$y = -4.$$

The y-intercept is $(0, -4)$. If $y = 0$,

$$x - y = 4$$
$$x - 0 = 4$$
$$x = 4.$$

The x-intercept is $(4, 0)$. The graph of the first line is

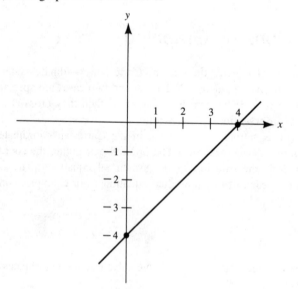

For the second line, if $x = 0$,

$$2x + 3y = 3$$
$$2(0) + 3y = 3$$
$$3y = 3$$
$$y = 1.$$

The y-intercept is $(0, 1)$. If $y = 0$,

$$2x + 3y = 3$$
$$2x + 3(0) = 3$$
$$2x = 3$$
$$x = \frac{3}{2}.$$

The x-intercept is $\left(\frac{3}{2}, 0\right)$.

We draw the graph of the second line on the same set of axes as the graph of the first line. The graph of the system is

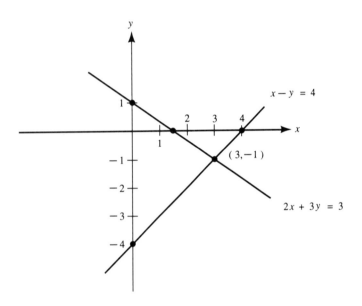

Observe that the point of intersection appears to have the coordinates $(3, -1)$.

We can find integer solutions of systems of equations by drawing the graph of the system and noting the coordinates of the point of intersection.

EXAMPLE 12.1   Graph the system

$$x + y - 5 = 0$$
$$x - 2y + 4 = 0$$

and find its solution.

Solution   First, we graph each of the two lines. As before, we draw graphs of lines by finding the y- and x-intercepts of each line. For the first line, the y-intercept is $(0, 5)$ and the x-intercept is $(5, 0)$. For the second line, the y-intercept is $(0, 2)$ and the x-intercept is $(-4, 0)$. The graph of the system is

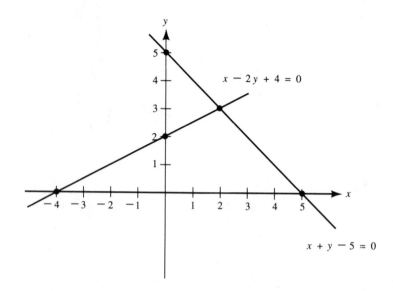

We see from the graph that the lines appear to intersect at the point (2, 3). To check, we substitute (2, 3) into each of the equations. For $x + y - 5 = 0$,

$$2 + 3 - 5 = 0,$$

and for $x - 2y + 4 = 0$,

$$2 - 2(3) + 4 = 0.$$

Therefore, (2, 3) is on both lines. Since this point is on both lines, it must be the point of the intersection. The ordered pair (2, 3) is the solution of the system.

It is a common error to confuse the words "intercept" and "intersect." The *intercepts* of each line are the points where each line crosses the y-axis and the x-axis. The lines *intersect* at the point where they cross one another.

EXAMPLE 12.2 Graph the system

$$3x + y - 3 = 0$$
$$2x - y + 8 = 0$$

and find its solution.

Solution   Using the intercepts to graph each line, we draw the graph of the system at the top of page 171. The point of intersection appears to be $(-1, 6)$. To check, we substitute in each equation. For $3x + y - 3 = 0$,

$$3(-1) + 6 - 3 \stackrel{?}{=} 0$$
$$-3 + 6 - 3 = 0.$$

For $2x - y + 8 = 0$,

$$2(-1) - 6 + 8 \stackrel{?}{=} 0$$
$$-2 - 6 + 8 = 0.$$

Therefore, the solution of the system is $(-1, 6)$.

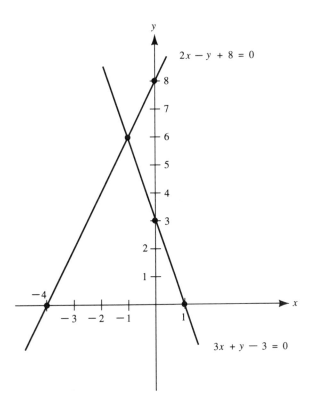

Graph for Example 12.2

The point of intersection of two lines can be the same as one of the intercepts.

EXAMPLE 12.3   Graph the system

$$3y - 2x = 6$$
$$2y - 3x = 9$$

and find its solution.

Solution   The graph of the system is

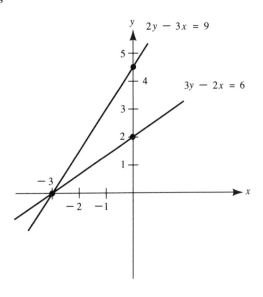

Each of the lines has x-intercept $(-3, 0)$. Clearly the point of intersection is $(-3, 0)$. Thus, $(-3, 0)$ is the solution of the system. You should check $(-3, 0)$ by substituting in each equation.

EXAMPLE 12.4   Graph the system

$$y = 4x + 6$$
$$y = 2$$

and find its solution.

Solution   The graph of the system is

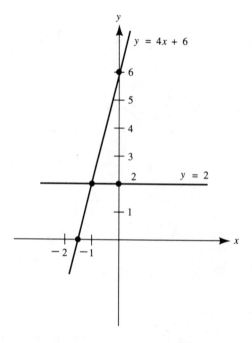

We see that the lines intersect at the point $(-1, 2)$. Thus, $(-1, 2)$ is the solution of the system. Clearly $(-1, 2)$ checks for the equation $y = 2$. You should check this solution by substituting in the first equation.

There are two special cases of systems of two linear equations in two variables. In one case, the lines are parallel.

EXAMPLE 12.5   Graph the system

$$2x + y = 3$$
$$2x + y + 5 = 0.$$

Solution   The graph of the system is

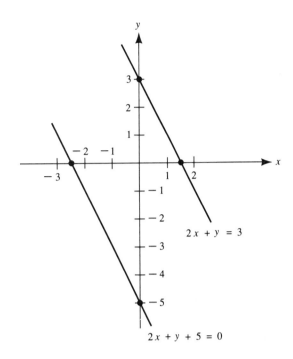

$$2x + y = 3$$

$$2x + y + 5 = 0$$

It appears that the lines are parallel. If this is the case, then the lines have no point of intersection, and the system has no solution. It is possible to show that the lines are indeed parallel. We recall the slope-intercept form of the equation, introduced in Section 10.5:

$$y = mx + b,$$

where m is the slope of the line and $(0, b)$ is the y-intercept. If we write the equations of the given lines in this form, we have

$$y = -2x + 3$$
$$y = -2x - 5.$$

When two lines have the same slope but different y-intercepts, then the lines are parallel. Here, both lines have slope $m = -2$. The y-intercept of the first line is $(0, 3)$ and the y-intercept of the second line is $(0, -5)$. Thus, the lines are parallel, there is no point of intersection, and the system has no solution. In this case, we say the system is **inconsistent.** An inconsistent system has no solution.

In the second special case, the lines are identical.

EXAMPLE 12.6   Graph the system

$$x - \frac{1}{2}y = 2$$
$$y = 2x - 4.$$

Solution   The graph of the system is

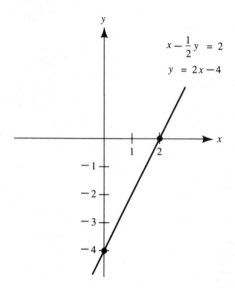

The lines are identical. Since each point of a pair of identical lines is a point of intersection, each point of the lines is a solution, and the system has infinitely many solutions. Writing the first equation in slope-intercept form, we have

$$x - \frac{1}{2}y = 2$$

$$2x - y = 4$$

$$2x - 4 = y$$

or

$$y = 2x - 4.$$

The second equation is already in slope-intercept form, and is identical to this result. In this case, we say the system is **dependent.** A dependent system has infinitely many solutions.

Exercise 12.1

Graph each system, and find the solution if it exists, or state that the system is inconsistent or dependent:

1. $x + y - 2 = 0$
   $x - y = 0$

2. $2x - y = 6$
   $x + 2y = -2$

3. $x - 3y = 3$
   $3y - 4x = 6$

4. $2x + 3y + 4 = 0$
   $4x + y - 2 = 0$

5. $3x + 4y = 12$
   $x - 2y = 4$

6. $3x - 2y = 6$
   $3x - 4y = 12$

7. $y = x + 2$
   $y = -1$

8. $x = -1$
   $x - y + 3 = 0$

9. $2x + y = 4$
   $2x + y = 2$

10. $2x - 3y = 6$
    $3y - 2x = 3$

11. $y = \frac{1}{2}x - 3$
    $x - 2y - 6 = 0$

12. $y = -3x + 3$
    $x + \frac{1}{3}y - 1 = 0$

12.2   The Addition Method

In the preceding section, we have assumed that the solutions of systems were ordered pairs of integers. Clearly, if the solutions involved fractions or decimal numbers, we would not be able to read the solutions accurately from graphs. For example, consider the system of equations

$$2x - y - 2 = 0$$
$$2x + 3y - 6 = 0.$$

The graph is

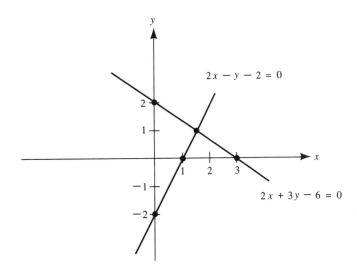

The x-coordinate of the point of intersection might be $\frac{3}{2}$, or $\frac{4}{3}$, or 1.4, or any of many other fractions or decimal numbers.

Using algebraic methods, we can find an exact solution of a system of equations. We eliminate one variable, and then can solve an equation in the remaining one variable. One way to eliminate a variable is the **addition method.**

Addition Rule: Given a system of equations, an equivalent system can be derived by adding like terms of two equations and replacing one of these two equations by the result.

EXAMPLE 12.7   Solve the system

$$x + y = 1$$
$$x - y = 5.$$

Solution   We add the two equations together by adding like terms on each side:

$$
\begin{array}{r}
x + y = 1 \\
x - y = 5 \\
\hline
2x \quad\;\; = 6 \\
x = 3.
\end{array}
$$

In the equation $x = 3$, we have eliminated the variable y. Now we may replace either one of the original equations by the equation $x = 3$. If we choose to replace the second equation, we have the equivalent system

$$x + y = 1$$
$$x = 3.$$

This system is easily solved. Replacing x by 3 in the first equation,

$$3 + y = 1$$
$$y = -2.$$

The solution is (3, −2). To check, we substitute (3, −2) in each equation. For $x + y = 1$,

$$3 + (-2) \overset{?}{=} 1$$
$$3 - 2 = 1,$$

and for $x - y = 5$,

$$3 - (-2) = 5$$
$$3 + 2 = 5.$$

EXAMPLE 12.8   Solve the system

$$2x - 3y - 5 = 0$$
$$x + 3y - 7 = 0.$$

Solution   We will generally write equations in the form

$$2x - 3y = 5$$
$$x + 3y = 7.$$

Now, we add the equations, and replace one of them by the result:

$$\begin{aligned} 2x - 3y &= 5 \\ x + 3y &= 7 \\ \hline 3x &= 12 \\ x &= 4. \end{aligned}$$

An equivalent system is

$$2x - 3y = 5$$
$$x = 4.$$

Replacing x by 4 in the first equation, we have

$$8 - 3y = 5$$
$$-3y = -3$$
$$y = 1.$$

Thus, the solution is (4, 1). You should check this solution by substituting in each of the original equations.

To solve the system given at the beginning of this section, we need an additional step.

EXAMPLE 12.9   Solve the system

$$2x - y - 2 = 0$$
$$2x + 3y - 6 = 0.$$

Solution   We write the system in the form

$$2x - y = 2$$
$$2x + 3y = 6.$$

Observe that if we simply add the equations together, we will not eliminate a variable. We must multiply both sides of the first equation by either -1 or 3. If we choose to multiply by -1, we have

$$-1(2x - y) = -1(2)$$
$$-2x + y = -2.$$

Now, we add the equations to eliminate one variable:

$$
\begin{array}{r}
-2x + y = -2 \\
2x + 3y = 6 \\
\hline
4y = 4 \\
y = 1.
\end{array}
$$

This result may replace one equation. An equivalent system is

$$2x - y = 2$$
$$y = 1.$$

Replacing y by 1 in the first equation,

$$2x - 1 = 2$$
$$2x = 3$$
$$x = \frac{3}{2}.$$

The solution is $\left(\frac{3}{2}, 1\right)$. To check, we substitute in the original equations. For $2x - y - 2 = 0$,

$$2\left(\frac{3}{2}\right) - 1 - 2 \overset{?}{=} 0$$
$$3 - 1 - 2 = 0.$$

For $2x + 3y - 6 = 0$,

$$2\left(\frac{3}{2}\right) + 3(1) - 6 \overset{?}{=} 0$$
$$3 + 3 - 6 = 0.$$

EXAMPLE 12.10   Solve the system

$$3x - y = -11$$
$$4x + 3y = 7.$$

Solution   If we multiply both sides of the first equation by 3, we can eliminate the y terms:

$$3(3x - y) = 3(-11)$$
$$9x - 3y = -33.$$

Now, we add the equations:

$$
\begin{array}{r}
9x - 3y = -33 \\
4x + 3y = 7 \\
\hline
13x = -26 \\
x = -2.
\end{array}
$$

An equivalent system is

$$3x - y = -11$$
$$x = -2.$$

Replacing x by -2 in the first equation,

$$3(-2) - y = -11$$
$$-6 - y = -11$$
$$-y = -5$$
$$y = 5.$$

The solution is $(-2, 5)$. You should check this solution by substituting in each of the original equations.

To solve some systems it is necessary to multiply each equation by an appropriate number. We must choose multipliers that will eliminate one of the variables when we add the equations.

EXAMPLE 12.11   Solve the system

$$3x - 4y = -6$$
$$4x - 3y = -1.$$

Solution   We will choose multipliers to make the y terms have the same absolute value but opposite signs. Since $3(-4y) = -12y$ and $-4(-3y) = 12y$, we may multiply both sides of the first equation by 3 and both sides of the second equation by -4:

$$3(3x - 4y) = 3(-6)$$
$$-4(4x - 3y) = -4(-1).$$

Now we remove the parentheses and add the equations:

$$9x - 12y = -18$$
$$-16x + 12y = 4$$
$$\overline{ }$$
$$-7x = -14$$
$$x = 2.$$

An equivalent system is

$$3x - 4y = -6$$
$$x = 2.$$

Replacing x by 2 in the first equation,

$$3(2) - 4y = -6$$
$$6 - 4y = -6$$
$$-4y = -12$$
$$y = 3.$$

The solution is $(2, 3)$. You should check this solution by substituting in each of the original equations.

We can identify systems that are inconsistent or dependent using algebraic methods.

EXAMPLE 12.12 Solve the system

$$3x - 2y = 1$$
$$6x - 4y = -1.$$

Solution We use the addition method, multiplying both sides of the first equation by -2:

$$-2(3x - 2y) = -2(1)$$
$$-6x + 4y = -2.$$

We add this result to the second equation:

$$-6x + 4y = -2$$
$$\underline{6x - 4y = -1}$$
$$0 = -3.$$

Since the statement $0 = -3$ is not true, there are no possible solutions. The system is inconsistent.

EXAMPLE 12.13 Solve the system

$$3x - 2y = -3$$
$$-\frac{3}{2}x + y = \frac{3}{2}.$$

Solution We use the addition method, multiplying both sides of the second equation by 2:

$$2\left(-\frac{3}{2}x + y\right) = 2\left(\frac{3}{2}\right)$$
$$-3x + 2y = 3.$$

We add the first equation to this result:

$$3x - 2y = -3$$
$$\underline{-3x + 2y = 3}$$
$$0 = 0.$$

Since the statement $0 = 0$ is true for all possible values of x and y, there are infinitely many solutions. The system is dependent.

Exercise 12.2

Solve the system:

1. $x + y = 8$
   $x - y = 2$

2. $x - y = -2$
   $x + y = 6$

3. $x - 2y - 7 = 0$
   $x + 2y - 3 = 0$

4. $x + 5y + 8 = 0$
   $x - 5y - 2 = 0$

5. $2x + 3y - 13 = 0$
   $x - 3y + 7 = 0$

6. $2x - 4y - 20 = 0$
   $5x + 4y - 8 = 0$

7. $3x - 3y = 5$
   $3x + 6y = 8$

8. $4x - 3y = 2$
   $4x + 2y = -8$

9. $x + 3y = 7$
   $2x + 3y = 5$

10. $2x + 2y = 5$
    $10x + 2y = 9$

11. $4x - y = 11$
    $3x + 5y = 14$

12. $5x - y = -11$
    $7x - 3y = -9$

13. $5x - 2y = 0$
    $-4x + 5y = 17$

14. $-3x + 4y = 0$
    $7x - 6y = 0$

15. $7x + 3y = 8$
    $3x + 5y = 9$

16. $6x - 3y + 11 = 0$
    $9x - 4y + 15 = 0$

17. $2x + y = 3$
    $4x + 2y = 5$

18. $6x - 9y - 2 = 0$
    $-4x + 6y + 3 = 0$

19. $-6x + 9y = 6$
    $10x - 15y = -10$

20. $x + \frac{1}{2}y - 2 = 0$
    $3x + \frac{3}{2}y - 6 = 0$

12.3   The Substitution Method

A second way to eliminate a variable is the **substitution method.** We use the substitution method when one of the equations has been solved for one variable, or can easily be solved for one variable.

> *Substitution Rule:*   Given a system of equations with one equation solved for one variable, an equivalent system can be derived by substituting for that variable in another equation, and replacing one of these two equations by the result.

EXAMPLE 12.14   Solve the system

$$x + y = 1$$
$$x = 2y + 7.$$

Solution   The second equation is solved for x. We substitute $2y + 7$ for x in the first equation:

$$x + y = 1$$
$$(2y + 7) + y = 1.$$

Then, we remove the parentheses, combine like terms, and solve for y:

$$2y + 7 + y = 1$$
$$3y + 7 = 1$$
$$3y = -6$$
$$y = -2.$$

We may replace either one of the original equations by this result. We form an equivalent system using the first equation with the equation $y = -2$:

$$x + y = 1$$
$$y = -2.$$

Then, replacing y by -2 in the first equation,

$$x + y = 1$$
$$x + (-2) = 1$$
$$x = 3.$$

The solution is $(3, -2)$. To check, we substitute in each of the original equations. For $x + y = 1$,

$$3 + (-2) = 1,$$

and for $x = 2y + 7$,

$$3 \stackrel{?}{=} 2(-2) + 7$$
$$3 = -4 + 7.$$

EXAMPLE 12.15   Solve the system

$$x = 3$$
$$2x + 4y = -3.$$

Solution   Since the first equation is solved for x, we simply replace x by 3 in the second equation:

$$2x + 4y = -3$$
$$2(3) + 4y = -3$$
$$6 + 4y = -3$$
$$4y = -9$$
$$y = -\frac{9}{4}.$$

The solution is $\left(3, -\frac{9}{4}\right)$. Clearly, $\left(3, -\frac{9}{4}\right)$ checks for $x = 3$. You should check this solution in the second equation.

If the variable we are substituting for is preceded by a constant or a negative, we must be sure to use parentheses and to remove them carefully.

EXAMPLE 12.16   Solve the system

$$y = 2x - 1$$
$$3x - y = 2.$$

Solution   The first equation is solved for y. We substitute $2x - 1$ for y in the second equation, being careful to use parentheses:

$$3x - y = 2$$
$$3x - (2x - 1) = 2$$
$$3x - 2x + 1 = 2$$
$$x + 1 = 2$$
$$x = 1.$$

An equivalent system is

$$y = 2x - 1$$
$$x = 1.$$

Replacing x by 1 in the first equation,

$$y = 2(1) - 1$$
$$y = 1.$$

The solution is $(1, 1)$. To check, we substitute in each of the original equations. For $y = 2x - 1$,

$$1 = 2(1) - 1.$$

For $3x - y = 2$,

$$3(1) - 1 = 2.$$

EXAMPLE 12.17   Solve the system

$$x = 4y + 10$$
$$2x - 3y = 5.$$

Solution   We substitute $4y + 10$ for x in the second equation, being careful to use parentheses:

$$2(4y + 10) - 3y = 5$$
$$8y + 20 - 3y = 5$$
$$5y + 20 = 5$$
$$5y = -15$$
$$y = -3.$$

An equivalent system is

$$x = 4y + 10$$
$$y = -3.$$

Replacing y by -3 in the first equation,

$$x = 4(-3) + 10$$
$$x = -2.$$

The solution is $(-2, -3)$. You should check this solution by substituting in each of the original equations.

EXAMPLE 12.18   Solve the system

$$2x - 3y - 3 = 0$$
$$y = \frac{2}{3}x - 1.$$

Solution   We substitute $\frac{2}{3}x - 1$ for y in the first equation:

$$2x - 3y - 3 = 0$$
$$2x - 3\left(\frac{2}{3}x - 1\right) - 3 = 0$$
$$2x - 2x + 3 - 3 = 0$$
$$0 = 0.$$

Since the statement $0 = 0$ is true for all possible values of x and y, there are infinitely many solutions. The system is dependent. If we had an untrue statement, there would be no solution and the system would be inconsistent.

Exercise 12.3

Solve the system:

1. $x + 2y = 6$
   $x = 3y - 4$

2. $x - y = 4$
   $x = 3y + 10$

3. $x + y = 7$
   $y = 2x - 5$

4. $5x + y = 1$
   $y = x - 2$

5. $x = -2$
   $3x + 4y = 4$

6. $2x - 3y = 12$
   $y = 4$

7. $y = 4x - 5$
   $x - y = 8$

8. $x = 2y - 3$
   $3y - x = 1$

9. $x = 3y + 8$
   $3x - y = 8$

10. $x = 3 - 2y$
    $2x + 3y = 5$

11. $y = 4 - x$
    $3x + 2y = 6$

12. $y = 3x + 1$
    $6x - 3y + 4 = 0$

13. $y = 3x + 5$
    $y = 4x + 8$

14. $y = x + 4$
    $y = -3x + 2$

15. $y = \dfrac{2}{3}x - 3$
    $2x - 3y = 9$

16. $y = \dfrac{5}{2}x + \dfrac{7}{4}$
    $10x - 4y + 5 = 0$

12.4   Applications

In many types of applications it is convenient, or necessary, to use two variables. For such applications, we use a system of two equations.

EXAMPLE 12.19   The sum of two numbers is 45, and their difference is 33. Find the numbers.

Solution   Let x and y represent the numbers. Then

$$x + y = 45$$

and

$$x - y = 33.$$

We solve the system of equations by addition:

$$
\begin{array}{rcl}
x + y &=& 45 \\
x - y &=& 33 \\
\hline
2x &=& 78 \\
x &=& 39.
\end{array}
$$

Since $x + y = 45$,

$$39 + y = 45$$
$$y = 6.$$

The numbers are 39 and 6. To check, we observe that $39 + 6 = 45$ and $39 - 6 = 33$.

It is possible to solve the preceding example using one variable. However, it is convenient to use two variables. Mixture problems are another type of problem where it is often convenient to use two variables. The following problem was solved using one variable in Section 8.3.

EXAMPLE 12.20   A 10-pound bag of potatoes contains Idahos, which sell for 45¢ a pound, and all-purpose Maines, which sell for 33¢ a pound. The bag costs $3.66. How many pounds of each kind of potato does it contain?

Solution   We let x represent the number of pounds of Idaho potatoes and y represent the number of pounds of Maine potatoes. We may make a chart as before:

	Unit price	Number of units	Total price
Idaho	45	x	$45x$
Maine	33	y	$33y$
Mixture		10	366

The sums of the second and third columns give the system of equations

$$x + y = 10$$
$$45x + 33y = 366.$$

Multiplying the first equation by -33 and adding,

$$
\begin{array}{r}
-33x - 33y = -330 \\
\underline{45x + 33y = 366} \\
12x = 36 \\
x = 3.
\end{array}
$$

Since $x + y = 10$,

$$3 + y = 10$$
$$y = 7.$$

There are three pounds of Idaho potatoes and seven pounds of Maine potatoes. You should check this solution.

We can also use systems of equations to solve a kind of rate, time, and distance problem that we cannot solve using only one variable.

EXAMPLE 12.21   An excursion boat cruises downstream for 3 hours. The passengers take the train back, but the crew spends 9 hours taking the boat back upstream. If the boat travels 45 miles each way, what are the average rates of the boat and the current?

Solution   We let r be the rate of the boat, and c be the rate of the current. Going downstream, the rates combine, and the actual rate is $r + c$. Going upstream, the current works against the boat, and the actual rate is $r - c$. We make the chart:

	Rate	Time	Distance
Downstream	$r + c$	3	45
Upstream	$r - c$	9	45

Since $RT = D$, we have the system of equations

$$(r + c)(3) = 45$$
$$(r - c)(9) = 45,$$

or, dividing the first equation by 3 and the second by 9,

$$r + c = 15$$
$$\underline{r - c = 5}$$
$$2r = 20$$
$$r = 10.$$

Since $r + c = 15$,

$$10 + c = 15$$
$$c = 5.$$

The average rate of the boat is 10 miles per hour, and the average rate of the current is 5 miles per hour. You should check this solution by finding the actual rate each way, and then showing that the distance each way is 45.

Exercise 12.4

1. The sum of two numbers is 64, and their difference is 28. Find the numbers.

2. The sum of two numbers is 25, and their difference is 8. Find the numbers.

3. Corn and lima beans are mixed to make a pound (16 ounces) of succotash. The corn costs 3¢ an ounce and the lima beans cost 5¢ an ounce. The pound of succotash costs 59¢. How many ounces of each vegetable does it contain?

4. You are paid $10.20 for 50 pounds of recycled bottles and cans. Bottles are worth 24¢ a pound and cans are worth 12¢ a pound. How many pounds of each did you recycle?

5. One quart (32 ounces) of a salad dressing base is made from oil and wine vinegar. The oil costs 8¢ an ounce and the wine vinegar costs $3\frac{1}{2}$¢ an ounce. If the quart of salad dressing base costs $2.20, how many ounces of oil and how many ounces of wine vinegar does it contain?

6. Herbs are planted in a mixture of potting soil and sand. If the potting soil costs 20¢ a pound and the sand costs 5¢ a pound, then the mixture costs 14¢ a pound. What fraction of a pound of the mixture is potting soil and what fraction is sand?

7. A canoeist paddles downstream 4 miles in 1 hour, but takes 2 hours to return the 4 miles upstream. What are the average rates of the canoe and the current?

8. A boat cruises downstream for $2\frac{1}{2}$ hours, but takes $12\frac{1}{2}$ hours to return upstream. The distance each way is 25 miles. What are the average rates of the boat and the current?

9. The distance between New York and Chicago is about 800 miles. A plane is scheduled to fly from New York to Chicago against the wind in 2 hours, but to return with the wind in $1\frac{2}{3}$ hours. What are the average rates of the plane and the wind expected to be?

10. A Rhine River steamer is scheduled to cruise upstream from Cologne to Mainz in $16\frac{2}{3}$ hours, and downstream from Mainz to Cologne in 10 hours. The distance between the cities is 200 kilometers. What are the average rates of the steamer and current expected to be?

Self-test

1. Solve by graphing:

    $x - 2y - 8 = 0$
    $x + y + 1 = 0$

2. Solve by addition:

    $3x + 4y = 9$
    $2x + 2y = 3$

    2. _____

3. Solve by substitution:

    $5x - y + 6 = 0$
    $y = 3x - 2$

    3. _____

4. Solve by addition or by substitution:

    $x + 5y = -2$
    $4x - 2y = 3$

    4. _____

5. An airplane takes 2 hours to fly 500 miles with the wind and $2\frac{1}{2}$ hours to return against the wind. Find the rate of the airplane and the rate of the wind.

    5. _____

UNIT

Linear Inequalities

INTRODUCTION

In preceding units you have solved equations, in which two expressions are equal. You have also drawn graphs of equations in two variables. In this unit you will learn how to solve inequalities, where two expressions are related so that one is less than or greater than the other. You will also learn how to draw graphs of inequalities in two variables.

OBJECTIVES

When you have finished this unit you should be able to:

1. Solve linear inequalities in one variable.
2. Draw graphs of linear inequalities in two variables.

13.1   Solving Linear Inequalities

Recall the inequality symbols from Section 3.2. The inequality symbols are $<$, "less than," and $>$, "greater than." We use these symbols, with equality sometimes included, to compare numbers in the following ways:

$a < b$ means a is less than b.
$a \leq b$ means a is less than or equal to b.
$a > b$ means a is greater than b.
$a \geq b$ means a is greater than or equal to b.

We often use the inequality symbols to represent sets of real numbers. For example,

$$x < 2$$

represents all real numbers less than 2, not including 2. However,

$$x \leq 2$$

represents all real numbers less than 2 and also including 2. Similarly,

$$x > 2$$

represents all real numbers greater than 2, not including 2, and

$$x \geq 2$$

represents all real numbers greater than 2 and including 2.

An **inequality** consists of two algebraic expressions related by an inequality symbol. The **solution of an inequality** is a set of real numbers such as those described above. In general, an inequality has infinitely many real number solutions.

To solve an inequality, we may add, subtract, multiply, and divide on both sides by the same number or expression. We use the same procedures as those in Unit 2, but with one important exception. Consider the following numerical example:

$$2 < 4$$

$$(-3)(2) \overset{?}{<} (-3)(4)$$

$$-6 \not< -12.$$

In fact,

$$-6 > -12.$$

In this example, we have multiplied both sides of an inequality by a negative number. The inequality symbol reverses from "less than" to "greater than." The direction of the inequality symbol is called the **sense of the inequality.** When we multiply an inequality by a negative number, the sense of the inequality reverses.

Now, consider this example of division of an inequality by a negative number:

$$2 < 4$$

$$\frac{2}{-1} \overset{?}{<} \frac{4}{-1}$$

$$-2 \not< -4.$$

In fact,

$$-2 > -4.$$

When we divide an inequality by a negative number, the sense of the inequality reverses.

Similar examples would indicate that, if we start with the inequality symbol "greater than" and multiply or divide by a negative number, the sense would reverse to the inequality symbol "less than."

Rule for Multiplying or Dividing Inequalities: When both sides of an inequality are multiplied or divided by a negative number, the sense of the inequality reverses.

EXAMPLE 13.1   Solve $x + 3 < 2$.

Solution   We subtract 3 from both sides of the inequality:

$$x + 3 < 2$$

$$x + 3 - 3 < 2 - 3$$

$$x < -1.$$

The sense of the inequality does not change when we add or subtract on both sides.

EXAMPLE 13.2   Solve $\frac{1}{3}x < 4$.

Solution   We multiply both sides of the inequality by 3:

$$\frac{1}{3}x < 4$$

$$3\left(\frac{1}{3}x\right) < 3(4)$$

$$x < 12.$$

The sense of the inequality does not change when we multiply or divide both sides by a positive number.

EXAMPLE 13.3   Solve $-2x < 8$.

Solution   We divide both sides of the inequality by -2:

$$-2x < 8$$

$$\frac{-2x}{-2} > \frac{8}{-2}$$

$$x > -4.$$

The sense of the inequality *does reverse* when we multiply or divide each side by a *negative* number.

We solve general linear inequalities exactly as we did general linear equations. We simplify any algebraic expressions, combine any like terms, and then solve the resulting inequality. However, when we multiply or divide both sides of an inequality by a negative number, we must remember to reverse the sense of the inequality.

EXAMPLE 13.4   Solve $2x - 4 < 6$.

Solution   We may add 4 to both sides of the inequality without changing its sense:

$$2x - 4 < 6$$

$$2x < 10.$$

Also, since 2 is positive, we may divide both sides of the inequality by 2 without changing its sense:

$$\frac{2x}{2} < \frac{10}{2}$$

$$x < 5.$$

All real numbers less than 5 are solutions. We call 5 the **boundary.**

We may graph this solution on a number line. We indicate the boundary, in this case 5, by an open dot when it is not included in the solution. The part of the line representing the numbers less than 5 is made heavier to indicate the solution $x < 5$:

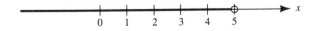

To check this solution, we first check that 5 is the boundary; that is, if we had an equation,

$$2x - 4 = 6$$
$$2(5) - 4 \overset{?}{=} 6$$
$$10 - 4 = 6.$$

Now, we check that the sense of the solution gives the correct side of the boundary. We choose any number less than 5 to check the inequality. For example, if $x = 4$,

$$2x - 4 < 6$$
$$2(4) - 4 \overset{?}{<} 6$$
$$8 - 4 \overset{?}{<} 6$$
$$4 < 6.$$

It is important to check both the boundary and the sense of the solution of an inequality.

EXAMPLE 13.5   Solve $6 - 3x < 9$.

Solution   We may subtract 6 from both sides of the inequality without changing the sense:

$$6 - 3x < 9$$
$$-3x < 3.$$

However, when we divide by the negative number -3, we must change the sense of the inequality:

$$-3x < 3$$
$$\frac{-3x}{-3} > \frac{3}{-3}$$
$$x > -1.$$

The solution is $x > -1$. The graph of this solution is

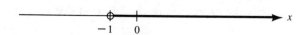

To check the boundary, we substitute -1 for x, as if we had an equation:

$$6 - 3x = 9$$
$$6 - 3(-1) \overset{?}{=} 9$$
$$6 + 3 = 9.$$

To check the sense, we may choose any number greater than -1 to check the inequality. Since $0 > -1$,

$$6 - 3x < 9$$
$$6 - 3(0) \overset{?}{<} 9$$
$$6 - 0 < 9.$$

Observe that we must check in the original inequality using the original sense.

An inequality may include equality. Then, the boundary is included in the solution.

EXAMPLE 13.6 Solve $6x - 2 \geq 7$.

Solution We add 2 to both sides of the inequality, and then divide both sides by 6. Neither of these operations changes the sense of the inequality:

$$6x - 2 \geq 7$$
$$6x \geq 9$$
$$x \geq \frac{9}{6}$$
$$x \geq \frac{3}{2}.$$

All real numbers greater than $\frac{3}{2}$, and also $\frac{3}{2}$, are solutions. Thus, the boundary is included in the solution.

We use a graph similar to those in the preceding examples to illustrate a solution which includes equality. The boundary $\frac{3}{2}$ is indicated by a solid dot when it is included in the solution. The solution $x \geq \frac{3}{2}$ may be illustrated by the graph

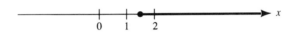

You should check the boundary and the sense of the solution in the original inequality.

If possible, we simplify the expressions on each side of the inequality before proceeding with the solution.

EXAMPLE 13.7 Solve $3(x - 2) - 4x \leq 2$.

Solution First, we simplify the expression on the left-hand side of the inequality:

$$3(x - 2) - 4x \leq 2$$
$$3x - 6 - 4x \leq 2$$
$$-x - 6 \leq 2.$$

Now, we add 6 to both sides, which does not change the sense of the inequality:

$$-x - 6 \leq 2$$
$$-x \leq 8.$$

Finally, we must multiply both sides by -1. This operation does change the sense of the inequality:

$$-x \leq 8$$
$$(-1)(-x) \geq (-1)(8)$$
$$x \geq -8.$$

The solution is $x \geq -8$. The graph is

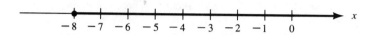

To check, we use the original expression. The boundary is -8:

$$3(x - 2) - 4x = 2$$
$$3(-8 - 2) - 4(-8) \stackrel{?}{=} 2$$
$$3(-10) + 32 \stackrel{?}{=} 2$$
$$-30 + 32 = 2.$$

To check the sense, we may use any number greater than -8. Since $0 > -8$,

$$3(x - 2) - 4x \leq 2$$
$$3(0 - 2) - 4(0) \stackrel{?}{\leq} 2$$
$$3(-2) - 0 \stackrel{?}{\leq} 2$$
$$-6 \leq 2.$$

Often we can avoid multiplying or dividing by a negative number by carefully choosing how we collect the x terms.

EXAMPLE 13.8   Solve $x - (2x + 2) > 3(x - 2)$.

Solution

First, we simplify the expressions on each side of the inequality:

$$x - (2x + 2) > 3(x - 2)$$
$$x - 2x - 2 > 3x - 6$$
$$-x - 2 > 3x - 6.$$

Now, we may avoid a negative coefficient in the x term by adding x to both sides:

$$-x - 2 > 3x - 6$$
$$-2 > 4x - 6.$$

Then, adding 6 to both sides and dividing by 4 does not change the sense of the inequality:

$$-2 > 4x - 6$$
$$4 > 4x$$
$$1 > x.$$

The solution $1 > x$ is the same as $x < 1$. We will usually write such solutions in the form $x < 1$. The graph is

You should check the boundary and sense of the solution using the original expressions.

☐ **Exercise 13.1**

Solve for x:

1. $x + 2 < 6$

2. $x - 9 > -5$

3. $\frac{1}{2}x > 5$

4. $6x < 12$

5. $-4x < -12$

6. $\frac{x}{-5} > 3$

7. $3x + 2 > 5$

8. $4x - 3 < 7$

9. $9 - 2x > 3$

10. $3 - 4x < 15$

11. $3x - 3 \le -4$

12. $1 - 9x \le -5$

13. $\frac{1}{4}x + 1 < \frac{5}{4}$

14. $\frac{1}{6} - \frac{1}{3}x \ge \frac{1}{2}$

15. $2(x + 1) - 4x > 1$

16. $2(x - 3) - 5x < 0$

17. $2x - (x - 3) < 3x + 2$

18. $2(x + 5) < 3x - (2 + 4x)$

19. $3(x - 2) - 3 \ge 6 - 4(x + 2)$

20. $5(x - 4) + 5 \le 4(2x - 3) - 3$

☐ **13.2** **Linear Inequalities in Two Variables**

Recall from Unit 10 that the solutions of a linear equation in two variables consist of infinitely many ordered pairs. We indicate such solutions by a graph. Similarly, the solutions of a linear inequality in two variables consist of infinitely many ordered pairs, which we indicate by a graph.

☐ **EXAMPLE 13.9**   Draw the graph of $x - 2y < 4$.

Solution   First, we draw the graph of the line $x - 2y = 4$. As before, we will use the y- and x-intercepts. If $x = 0$,

$$0 - 2y = 4$$
$$-2y = 4$$
$$y = -2.$$

The y-intercept is $(0, -2)$. If $y = 0$,

$$x - 2(0) = 4$$
$$x - 0 = 4$$
$$x = 4.$$

The *x*-intercept is (4, 0). When we draw the line, we use a dashed line. Equality is not included in the given inequality, so the line is not actually part of the graph of solutions. The dashed line is the boundary of the graph:

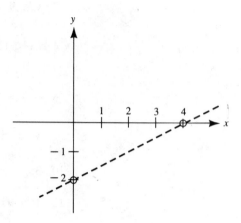

Now, we solve the inequality $x - 2y < 4$ for y:

$$x - 2y < 4$$
$$-2y < 4 - x.$$

We must divide by -2, so the sense of the inequality reverses:

$$\frac{-2y}{-2} > \frac{4 - x}{-2}$$
$$y > \frac{4 - x}{-2}.$$

When an inequality is solved for y, and y is greater than the expression involving x, the graph consists of the points above the line. We indicate the graph by shading this area:

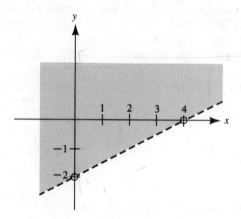

To check, we choose any point in the shaded area of the graph. In this case, the origin (0, 0) is in the shaded area. We substitute (0, 0) in the original inequality:

$$x - 2y < 4$$

$$0 - 2(0) \overset{?}{<} 4$$

$$0 - 0 \overset{?}{<} 4$$

$$0 < 4.$$

When an inequality includes equality, the boundary is included in the graph.

EXAMPLE 13.10   Draw the graph of $3x + 6y \le 9$.

Solution   We draw the graph of $3x + 6y = 9$, using a solid line, because the boundary is included. Then, solving for y,

$$3x + 6y \le 9$$

$$6y \le 9 - 3x$$

$$y \le \frac{9 - 3x}{6},$$

where the sense of the inequality does not reverse in dividing by 6. When an inequality is solved for y, and y is less than the expression involving x, the graph consists of the points below the line. We indicate the graph by shading this area:

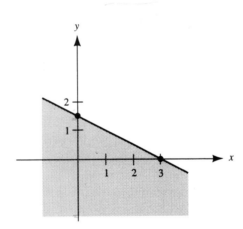

You should check this solution using any point, such as $(0, 0)$, in the shaded part of the graph.

EXAMPLE 13.11   Draw the graph of $4x - y - 4 \ge 0$.

Solution   We draw the graph of $4x - y - 4 = 0$, using a solid line. Then, solving for y,

$$4x - y - 4 \ge 0$$

$$4x - 4 \ge y$$

or

$$y \le 4x - 4.$$

We could also solve by writing

$$4x - y - 4 \geq 0$$
$$-y \geq -4x + 4,$$

and, multiplying by -1 and reversing the sense of the inequality,

$$(-1)(-y) \leq (-1)(-4x + 4)$$
$$y \leq 4x - 4.$$

Since y is less than the expression involving x, we shade the part below the line:

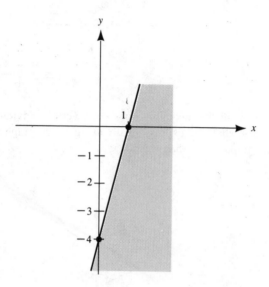

To check, we observe that $(0, 0)$ is *not* in the shaded part. Thus, $(0, 0)$ should cause the original inequality to be false:

$$4x - y - 4 \geq 0$$
$$4(0) - 0 - 4 \overset{?}{\geq} 0$$
$$0 - 0 - 4 \overset{?}{\geq} 0$$
$$-4 \ngeq 0.$$

Alternatively, we may choose any point in the shaded part, and check that it makes the inequality true. For example, $(2, 0)$ is in the shaded part. Substituting $(2, 0)$ in the inequality,

$$4x - y - 4 \geq 0$$
$$4(2) - 0 - 4 \overset{?}{\geq} 0$$
$$8 - 0 - 4 \overset{?}{\geq} 0$$
$$4 \geq 0.$$

EXAMPLE 13.12   Draw the graph of $y > x$.

Solution   We draw the graph of $y = x$, using a dashed line because equality is not included. Since both intercepts are $(0, 0)$, we must use some other point such as $(1, 1)$. Then, since the inequality is already solved for y and $y > x$, we shade the part above the line:

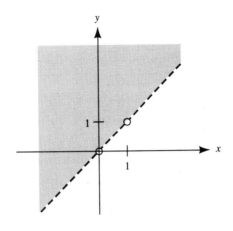

Observe that the graph of $y > x$ includes exactly those points for which the y-coordinate is greater than the x-coordinate.

Exercise 13.2

Draw the graph:

1. $y < x + 2$ 2. $y \leq 2x - 1$ 3. $y > 2x - 4$ 4. $y \geq 5 - 2x$

5. $x - 2y < 2$ 6. $2x - 5y < 10$ 7. $2x + 3y \leq 6$ 8. $3y - 4x \leq 12$

9. $3x - 4y + 6 \geq 0$ 10. $5x - 2y - 5 \geq 0$ 11. $3x - 4y > 12$ 12. $x + 2y > 4$

13. $y > -\dfrac{1}{2}x$ 14. $y \leq \dfrac{2}{3}x$ 15. $y - 2 < 0$ 16. $x + 3 \geq 0$

Self-test

Solve and check:

1. $3(x - 6) < x - (2x - 2)$

2. $9 - 6x \leq 2x - 15$

3. $9x + 23 > 2$

Draw the graph:

4. $2x - 3y + 12 < 0$

5. $y \geq 3x - 4$

1. _____

2. _____

3. _____

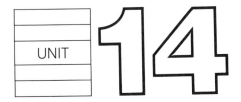

UNIT 14

Exponents

INTRODUCTION

In most of the preceding units you have used linear expressions. Occasionally, you used expressions that involved products of two or more variables or reciprocals of variables. Such expressions are not linear. In this unit you will learn about another type of nonlinear expression. These expressions involve exponents that are positive integers, such as squares or cubes of variables. You will learn how to evaluate such expressions, their basic rules, and how to simplify them. Then, you will learn an application of exponents called scientific notation.

OBJECTIVES

When you have finished this unit you should be able to:

1. Find the value of a power of a positive or negative integer.
2. Evaluate expressions involving powers of positive or negative integers.
3. Use the rules for exponents to simplify expressions involving positive integral exponents.
4. Write numbers in scientific notation, and multiply numbers given in scientific notation.

14.1 Powers of Integers

You may remember from other courses that, for example, 3^2, or "three squared," means $(3)(3) = 9$. More generally,

$$x^2 = x \cdot x \text{ (read "}x\text{ squared")}$$

means two factors of x. Similarly,

$$x^3 = x \cdot x \cdot x \text{ (read "}x\text{ cubed")}$$

means three factors of x.

$$x^4 = x \cdot x \cdot x \cdot x \text{ (read "}x\text{ to the fourth")}$$

means four factors of x. In general, an expression of the form b^n is called the **nth power of b.** In the expression b^n, b is called the **base** and n is called the **exponent.**

Definition: $b^n = \underbrace{b \cdot b \cdot b \cdot \ldots \cdot b}_{n \text{ factors}}$, or n factors of b.

This definition makes sense only for exponents n that are positive integers: 1, 2, 3, 4, and so on. In particular, observe that $b^1 = b$, one factor of b.

EXAMPLE 14.1   Find the value:

a. 4^3 b. 3^4 c. 5^1

Solutions

a. "Four cubed" is three factors of 4:

$$4^3 = (4)(4)(4) = 64.$$

b. "Three to the fourth" is four factors of 3:

$$3^4 = (3)(3)(3)(3) = 81.$$

c. "The first power of five" is one factor of 5:

$$5^1 = 5.$$

EXAMPLE 14.2   Find the value:

a. $(-5)^2$ b. $(-2)^5$ c. $(-2)^1$

Solutions

a. "Negative five squared" is two factors of -5:

$$(-5)^2 = (-5)(-5) = 25.$$

b. "Negative two to the fifth" is five factors of -2:

$$(-2)^5 = (-2)(-2)(-2)(-2)(-2) = -32.$$

c. "The first power of negative two" is one factor of -2:

$$(-2)^1 = -2.$$

If n is odd, b^n is called an **odd power**. If n is even, b^n is called an **even power**. Observe that an odd power of a negative number is negative, but an even power of a negative number is positive.

EXAMPLE 14.3   Find the value:

a. $(-6)^2$ b. $(-6)^3$

Solutions

a. "Negative six squared" is an even power:

$$(-6)^2 = (-6)(-6) = 36,$$

which is positive.

b. "Negative six cubed" is an odd power:

$$(-6)^3 = (-6)(-6)(-6) = -216,$$

which is negative.

We must be careful in determining the meaning of parentheses. When a minus sign is inside the parentheses, an exponent outside the parentheses applies to the minus sign. If a minus sign is outside the parentheses, or if there are no parentheses, an exponent does not apply to the minus sign.

EXAMPLE 14.4 Find the value:

a. $(-8)^2$ b. -8^2

Solutions   a. The minus sign is inside the parentheses. Therefore,

$$(-8)^2 = (-8)(-8) = 64.$$

b. There are no parentheses. Therefore,

$$-8^2 = -(8)(8) = -64.$$

EXAMPLE 14.5 Find the value of $-(-2)^2$.

Solution   The minus sign outside the parentheses is not affected by the exponent. The exponent does apply to the minus sign inside the parentheses. Therefore,

$$-(-2)^2 = -(-2)(-2) = -(4) = -4.$$

Since the product of any number of factors of 1 is 1, any power of 1 equals 1. Also, any even power of -1 equals 1. Any odd power of -1 equals -1.

EXAMPLE 14.6 Find the value:

a. 1^5 b. $(-1)^4$ c. $(-1)^3$

Solutions   a. Any power of 1 equals 1:

$$1^5 = (1)(1)(1)(1)(1) = 1.$$

b. An even power of -1 equals 1:

$$(-1)^4 = (-1)(-1)(-1)(-1) = 1.$$

c. An odd power of -1 equals -1:

$$(-1)^3 = (-1)(-1)(-1) = -1.$$

Finally, since the product of any number of factors of 0 is 0, any power of 0 equals 0.

Exercise 14.1

Find the value:

1. 5^3

2. 3^5

3. 6^1

4. $(-4)^1$

5. $(-2)^4$

6. $(-4)^3$

7. $(-5)^3$

8. $(-12)^2$

9. $(-7)^2$

10. -7^2

11. $(-3)^3$

12. -3^3

13. $-(-6)^2$ 14. $-(-2)^3$ 15. $-(-3)^5$ 16. $-(-5)^4$

17. $(-10)^3$ 18. $(-10)^6$ 19. $(-1)^5$ 20. $(-1)^8$

21. -1^4 22. $-(-1)^3$ 23. 0^5 24. 0^{10}

14.2 Expressions Involving Exponents

In this section we evaluate expressions involving exponents. This process is the same as evaluating other expressions. However, we must be careful in calculating the signs of powers.

EXAMPLE 14.7 Evaluate $x^2 + y^2$ if $x = 2$ and $y = -3$.

Solution   We substitute 2 for x and -3 for y:

$$x^2 + y^2 = 2^2 + (-3)^2$$
$$= 4 + 9$$
$$= 13.$$

EXAMPLE 14.8 Evaluate $x^2 - y^2$ if $x = 2$ and $y = -3$.

Solution   Again, we substitute 2 for x and -3 for y:

$$x^2 - y^2 = 2^2 - (-3)^2.$$

The first minus sign is a subtraction symbol outside of the parentheses and is not squared. The -3 is squared:

$$x^2 - y^2 = 2^2 - (-3)^2$$
$$= 4 - 9$$
$$= -5.$$

EXAMPLE 14.9 Evaluate $x^2 - y^3$ if $x = 2$ and $y = -3$.

Solution   The -3 is cubed, but the cube of a negative number is negative:

$$x^2 - y^3 = 2^2 - (-3)^3$$
$$= 4 - (-27)$$
$$= 4 + 27$$
$$= 31.$$

In later units we will deal with **quadratic expressions,** which contain the square of a variable, but no higher power.

EXAMPLE 14.10 Evaluate $2x^2 - 3x - 1$ if $x = -5$.

Solution   We substitute -5 for x wherever x appears:

$$2x^2 - 3x - 1 = 2(-5)^2 - 3(-5) - 1.$$

We must square -5 before multiplying by 2:

$$
\begin{aligned}
2x^2 - 3x - 1 &= 2(-5)^2 - 3(-5) - 1 \\
&= 2(25) + 15 - 1 \\
&= 50 + 15 - 1 \\
&= 65 - 1 \\
&= 64.
\end{aligned}
$$

EXAMPLE 14.11   Evaluate $-x^2 + 4x + 4$ if $x = 8$.

Solution   The square does not apply to the minus sign. When we substitute 8 for x, only 8 is squared:

$$
\begin{aligned}
-x^2 + 4x + 4 &= -8^2 + 4(8) + 4 \\
&= -64 + 32 + 4 \\
&= -32 + 4 \\
&= -28.
\end{aligned}
$$

An expression often used in dealing with quadratic expressions is $b^2 - 4ac$.

EXAMPLE 14.12   Evaluate $b^2 - 4ac$ if $a = 2$, $b = -5$, and $c = -4$.

Solution   We substitute -5 for b, 2 for a, and -4 for c, and calculate, taking care in placing parentheses:

$$
\begin{aligned}
b^2 - 4ac &= (-5)^2 - 4(2)(-4) \\
&= 25 - 4(-8) \\
&= 25 + 32 \\
&= 57.
\end{aligned}
$$

Exercise 14.2

Evaluate:

1. $x^2 + y^2$ if $x = 3$ and $y = -4$
2. $a^2 + b^2$ if $a = -5$ and $b = 2$
3. $x^2 - y^2$ if $x = 3$ and $y = -4$
4. $p^2 - q^2$ if $p = -8$ and $q = -2$
5. $x^2 - y^3$ if $x = 3$ and $y = -4$
6. $a^3 - b^3$ if $a = 3$ and $b = -2$
7. $p^3 + q^3$ if $p = -5$ and $q = -4$
8. $s^3 - t^3$ if $s = -1$ and $t = -2$
9. $2x^2 - 4x - 5$ if $x = -3$
10. $3x^2 - 10x + 3$ if $x = -1$
11. $-x^2 + 5x + 6$ if $x = -2$
12. $-2x^2 - x - 1$ if $x = -4$
13. $b^2 - 4ac$ if $a = 3$, $b = -4$, and $c = -1$
14. $b^2 - 4ac$ if $a = -2$, $b = -1$, and $c = -5$
15. $a^2 - ab + b^2$ if $a = 5$ and $b = -6$
16. $x^2 - 2xy - y^2$ if $x = -4$ and $y = -7$

14.3   Rules for Exponents

In this section we will state five basic rules for exponents. We will not attempt to prove these rules, but we will illustrate each rule by an example.

Rules of Exponents: For m and n any positive integers:

1. $b^m \cdot b^n = b^{m+n}$.

2. a. $\dfrac{b^m}{b^n} = b^{m-n}$ if $m > n$ (m larger than n), and $b \neq 0$.

   b. $\dfrac{b^m}{b^n} = \dfrac{1}{b^{n-m}}$ if $n > m$ (n larger than m), and $b \neq 0$.

   c. $\dfrac{b^n}{b^n} = 1$, $b \neq 0$.

3. $(b^m)^n = b^{mn}$.

4. $(ab)^n = a^n b^n$.

5. $\left(\dfrac{a}{b}\right)^n = \dfrac{a^n}{b^n}$, $b \neq 0$.

EXAMPLE 14.13   Simplify $x^4 \cdot x^3$.

Solution   Using Rule 1,

$$x^4 \cdot x^3 = x^{4+3}$$
$$= x^7.$$

Observe that

$$x^4 \cdot x^3 = (x \cdot x \cdot x \cdot x)(x \cdot x \cdot x),$$

or seven factors of x. Although such an example is not a proof of the rule, the example should help you to see why the rule works. To multiply two powers with the same base, we add the exponents.

It is important to remember that in using Rule 1, although we add the exponents, we are actually multiplying the powers. There are no rules for adding or subtracting powers. Expressions of the form $b^m + b^n$ or $b^m - b^n$ usually cannot be simplified.

EXAMPLE 14.14   Simplify $x^3 + x^2$.

Solution   Since x^3 and x^2 are not like terms, the expression cannot be simplified.

EXAMPLE 14.15   Simplify:

a. $\dfrac{x^5}{x^2}$ b. $\dfrac{x}{x^6}$ c. $\dfrac{x^3}{x^3}$

Solutions   a. Using Rule 2a,

$$\frac{x^5}{x^2} = x^{5-2}$$
$$= x^3.$$

Observe that

$$\frac{x^5}{x^2} = \frac{\cancel{x} \cdot \cancel{x} \cdot x \cdot x \cdot x}{\cancel{x} \cdot \cancel{x}} = x \cdot x \cdot x,$$

or three factors of x.

b. Using Rule 2b,

$$\frac{x}{x^6} = \frac{1}{x^{6-1}}$$
$$= \frac{1}{x^5}.$$

Observe that

$$\frac{x}{x^6} = \frac{\cancel{x}}{\cancel{x} \cdot x \cdot x \cdot x \cdot x \cdot x} = \frac{1}{x \cdot x \cdot x \cdot x \cdot x},$$

leaving five factors of x in the denominator. To divide powers with the same base but different exponents, we subtract the smaller exponent from the larger. The result is written in the numerator if the larger exponent is in the numerator, with the denominator understood to be 1. The result is written in the denominator if the larger exponent is in the denominator and the numerator is 1.

c. Using Rule 2c,

$$\frac{x^3}{x^3} = 1.$$

Observe that

$$\frac{x^3}{x^3} = \frac{\cancel{x} \cdot \cancel{x} \cdot \cancel{x}}{\cancel{x} \cdot \cancel{x} \cdot \cancel{x}} = \frac{1}{1} = 1.$$

EXAMPLE 14.16   Simplify $(x^3)^4$.

Solution   Using Rule 3,

$$(x^3)^4 = x^{3 \cdot 4}$$
$$= x^{12}.$$

Observe that,

$$(x^3)^4 = x^3 \cdot x^3 \cdot x^3 \cdot x^3 = (x \cdot x \cdot x)(x \cdot x \cdot x)(x \cdot x \cdot x)(x \cdot x \cdot x)$$

or twelve factors of x.

☐ **EXAMPLE 14.17**   Simplify $(xy)^3$.

Solution   Using Rule 4,

$$(xy)^3 = x^3 y^3.$$

Observe that

$$(xy)^3 = (xy)(xy)(xy) = (x \cdot x \cdot x)(y \cdot y \cdot y),$$

or the product of three factors of x and three factors of y.

☐ **EXAMPLE 14.18**   Simplify $\left(\dfrac{x}{y}\right)^5$.

Solution   Using Rule 5,

$$\left(\frac{x}{y}\right)^5 = \frac{x^5}{y^5}.$$

Observe that

$$\left(\frac{x}{y}\right)^5 = \frac{x}{y} \cdot \frac{x}{y} \cdot \frac{x}{y} \cdot \frac{x}{y} \cdot \frac{x}{y} = \frac{x \cdot x \cdot x \cdot x \cdot x}{y \cdot y \cdot y \cdot y \cdot y}$$

or the quotient of five factors of x and five factors of y.

We will often encounter expressions in which two or more of the rules for exponents are involved.

☐ **EXAMPLE 14.19**   Simplify $(pq^2)^3$.

Solution   First, we use Rule 4, with $a = p$ and $b = q^2$:

$$(pq^2)^3 = p^3 (q^2)^3.$$

The second factor now can be simplified using Rule 3:

$$p^3 (q^2)^3 = p^3 q^6.$$

Thus, we have

$$(pq^2)^3 = p^3 q^6.$$

You can often do examples of this type in one step. However, you must be careful to distinguish between examples of the form $(pq^2)^3$ and examples of the form $p(q^2)^3$. In the second expression, the parentheses indicate that the exponent 3 applies only to the factor q^2, and not to the factor p. Thus,

$$p(q^2)^3 = pq^6.$$

In the next two examples, the placement of parentheses is also important.

EXAMPLE 14.20   Simplify $\left(\dfrac{r^5 s t^2}{r s^3 t^2}\right)^4$.

Solution   First working inside the parentheses, we apply the different parts of Rule 2 to each pair of powers with the same base. Using Rule 2a gives

$$\frac{r^5}{r} = r^4 \text{ or } \frac{r^4}{1},$$

Rule 2b gives

$$\frac{s}{s^3} = \frac{1}{s^2},$$

and Rule 2c gives

$$\frac{t^2}{t^2} = 1.$$

Therefore,

$$\left(\frac{r^5 s t^2}{r s^3 t^2}\right)^4 = \left(\frac{r^4}{s^2}\right)^4.$$

Now, applying Rule 5,

$$\left(\frac{r^4}{s^2}\right)^4 = \frac{(r^4)^4}{(s^2)^4}.$$

Finally, by Rule 3,

$$\frac{(r^4)^4}{(s^2)^4} = \frac{r^{16}}{s^8}.$$

Usually, you can apply Rules 5 and 3 in the same step:

$$\left(\frac{r^5 s t^2}{r s^3 t^2}\right)^4 = \left(\frac{r^4}{s^2}\right)^4$$
$$= \frac{r^{16}}{s^8}.$$

EXAMPLE 14.21   Simplify $\dfrac{(r^2 s)^3}{(r s^2)^2}$.

Solution   In this example we cannot begin by applying Rule 2, because different exponents apply to the numerator and the denominator. Therefore, we use Rule 3 first:

$$\frac{(r^2 s)^3}{(r s^2)^2} = \frac{r^6 s^3}{r^2 s^4}$$
$$= \frac{r^4}{s}.$$

Exercise 14.3

Simplify:

1. x^3x^5

2. $x^{10}x^5$

3. $x^{10} + x^5$

4. $x^3 - x^4$

5. $\dfrac{x^6}{x^2}$

6. $\dfrac{x^2}{x^4}$

7. $\dfrac{x^2}{x^2}$

8. $\dfrac{x^{10}}{x^9}$

9. $(x^3)^6$

10. $(x^5)^2$

11. $(xy)^3$

12. $\left(\dfrac{x}{y}\right)^8$

13. $(p^2q^3)^4$

14. $(p^5q^6)^2$

15. $p^3(q^3)^5$

16. $(p^3q^3)^5$

17. $2r(s^2)^4$

18. $(2rs^2)^4$

19. $(5r^3s)^2$

20. $5(r^3s)^2$

21. $\left(\dfrac{rs^3}{st^3}\right)^2$

22. $\left(\dfrac{r^2s^3t^4}{r^4s^3t^2}\right)^3$

23. $\dfrac{(rs^3)^5}{(rs)^4}$

24. $\dfrac{(rst)^3}{(r^2s^3t^4)^2}$

14.4   Scientific Notation

Exponents offer a convenient way to write numbers, especially numbers that are very large, and numbers that are very close to zero. Because this way of writing numbers is often used in the sciences, it is called **scientific notation.** A positive number, in scientific notation, is written as the product of a number between 1 and 10 and a power of 10. These are examples of numbers written in scientific notation:

$$2.38 \times 10^2, \quad 5.6 \times 10^6, \quad 4.905 \times 10^{12}.$$

Observe that, in the first factor, the decimal point always follows the first digit, and the first digit is not zero. The multiplication symbol $\times$ is traditionally used in scientific notation. However, multiplication parentheses may be used:

$$2.38(10^2), \quad 5.6(10^6), \quad 4.905(10^{12}).$$

Recall from arithmetic that multiplication by 10 moves the decimal point one place to the right. With this rule in mind, we can convert numbers given in scientific notation into ordinary notation.

EXAMPLE 14.22   Write the number in ordinary notation:

a. 2.38×10^2 b. 5.6×10^6 c. 4.905×10^{12}

Solutions

a. We must multiply by 10 twice, so the decimal point moves two places to the right:

$$2.38 \times 10^2 = 238.$$

b. We multiply by six factors of 10, so the decimal point moves six places to the right. We must supply five zeroes:

$$5.6 \times 10^6 = 5,600,000.$$

c. We multiply by twelve factors of 10 and must supply nine zeroes:

$$4.905 \times 10^{12} = 4,905,000,000,000.$$

You can see why scientific notation is a convenient way to write large numbers. It is much easier to read and write 4.905×10^{12} than 4,905,000,000,000. Moreover, when you use scientific notation, you do not need to worry about miscounting and losing a zero.

To write a number larger than 10 in scientific notation, we first mark the place where the decimal point must go, after the first nonzero digit. We then count the number of digits from the mark to the end of the number, including zeroes. This number of places is our power of 10.

EXAMPLE 14.23   Write the number in scientific notation:

a. 2340   b. 91,360,000   c. 202,000,000,000

Solutions

a. We mark the place for the decimal point after the first nonzero digit:

$$2_{\wedge}340.$$

The mark we have used is called a caret. Now, we count three digits from the caret to the end of the number. The power of 10 is 10^3. Therefore, to write the number in scientific notation, we put the decimal point where the caret is and multiply by 10^3:

$$2340 = 2.34 \times 10^3.$$

After we have accounted for all places, we may drop any zeroes following the last nonzero digit.

b. We use the caret to mark the place for the decimal point:

$$9_{\wedge}1,360,000.$$

Then, since there are seven digits following the caret, the power of 10 is 10^7. Therefore, in scientific notation,

$$91,360,000 = 9.136 \times 10^7.$$

c. We use the caret to mark the place for the decimal point:

$$2_{\wedge}02,000,000,000.$$

Then, since there are 11 digits following the caret, the power of 10 is 10^{11}. Therefore, in scientific notation,

$$202,000,000,000 = 2.02 \times 10^{11}.$$

Of course, we may not drop the zero between the twos.

Now, we consider numbers that are very close to zero, such as 0.00234. Observe the numbers in this list:

$$2340 = 2.34 \times 10^3$$
$$234 = 2.34 \times 10^2$$
$$23.4 = 2.34 \times 10^1$$
$$2.34 = 2.34 \times 10^0$$
$$0.234 = 2.34 \times 10^{-1}$$
$$0.0234 = 2.34 \times 10^{-2}$$
$$0.00234 = 2.34 \times 10^{-3}$$

We have not previously considered zero or negative exponents. Such exponents are covered in more advanced algebra texts. For the purposes of this book, we will simply state a meaning for powers of 10 with zero or negative exponents. We will take 10^0 (ten to the zero) to mean *no* factors of 10, that is, no multiplication by 10. Therefore,

$$2.34 = 2.34 \times 10^0.$$

We will take 10^{-1} (ten to the negative one) to mean *division* by 10. Recall that division by 10 moves the decimal point one place to the left. Therefore,

$$0.234 = 2.34 \times 10^{-1}.$$

Similarly, 10^{-2} means division by two factors of 10, 10^{-3} means division by three factors of 10, and so on.

EXAMPLE 14.24   Write the number in ordinary notation:

a. 4.1×10^{-4} b. 3.06×10^{-10} c. 9.5×10^0.

Solutions

a. We divide by four factors of 10, so the decimal point moves four places to the left. We must supply three zeroes between the decimal point and the 4:

$$4.1 \times 10^{-4} = 0.00041.$$

b. We divide by ten factors of 10, so the decimal point moves ten places to the left. We must supply nine zeroes between the decimal point and the 3:

$$3.06 \times 10^{-10} = 0.000000000306.$$

c. We do not multiply or divide by any factors of 10:

$$9.5 \times 10^0 = 9.5.$$

To write a number close to zero in scientific notation, we mark the place where the decimal point must go as before. Again, this place is after the first nonzero digit. Then, we count the number of digits, including zeroes, going left to the actual location of the decimal point. The negative of this number is the exponent for the power of 10.

EXAMPLE 14.25   Write the number in scientific notation:

a. 0.0155   b. 0.00003104   c. 0.0000000000002

Solutions

a. We use the caret to mark the place for the decimal point, after the first nonzero digit:

$$0.01_\wedge 55.$$

Now, we count two digits to the left to the actual decimal point. The power of 10 is 10^{-2}. Therefore, in scientific notation,

$$0.0155 = 1.55 \times 10^{-2}.$$

b. We use the caret to mark the place for the decimal point:

$$0.00003_\wedge 104.$$

Then, since there are five digits to the left to the actual decimal point, the power of 10 is 10^{-5}. Therefore, in scientific notation

$$0.00003104 = 3.104 \times 10^{-5}.$$

c. We use the caret to mark the place for the decimal point:

$$0.0000000000002_\wedge$$

Then, since there are 13 digits to the left to the actual decimal point, the power of 10 is 10^{-13}. Therefore, in scientific notation,

$$0.0000000000002 = 2 \times 10^{-13}.$$

To multiply two numbers written in scientific notation, we multiply their first factors and, separately, multiply the powers of 10 using Rule 1 for exponents.

EXAMPLE 14.26   Multiply $(5.3 \times 10^8)(4.9 \times 10^9)$, and write the result in scientific notation.

Solution   We calculate $(5.3)(4.9)$ to obtain 25.97. Then, using Rule 1 for exponents,

$$(10^8)(10^9) = 10^{8+9}$$
$$= 10^{17}.$$

Therefore,

$$(5.3 \times 10^8)(4.9 \times 10^9) = 25.97 \times 10^{17}.$$

However, the result is not in scientific notation. The decimal point should be after the 2:

$$2_\wedge 5.97 \times 10^{17}.$$

We count one place to the current location of the decimal point. Thus, we need one more factor of 10:

$$25.97 \times 10^{17} = 2.597 \times 10 \times 10^{17}$$
$$= 2.597 \times 10^{18}.$$

Therefore,

$$(5.3 \times 10^8)(4.9 \times 10^9) = 2.597 \times 10^{18},$$

written in scientific notation.

There are many applications, involving large numbers or numbers very near zero, in which it is advantageous to use numbers written in scientific notation.

EXAMPLE 14.27 The mass of a proton is 1836 times the mass of an electron. The mass of an electron is 0.000000000000000000000000000911 grams. What is the mass of a proton in scientific notation, to two decimal places?

Solution   Clearly, it is advantageous to write the mass of an electron in scientific notation:

$$0.000000000000000000000000000911 = 9.11 \times 10^{-28}.$$

We may also write 1836 in scientific notation:

$$1836 = 1.836 \times 10^3.$$

Then we multiply, assuming that the rules for exponents apply to negative exponents:

$$
\begin{aligned}
(1.836 \times 10^3)(9.11 \times 10^{-28}) &= (1.836)(9.11) \times 10^{3+(-28)} \\
&= 16.7 \times 10^{-25} \\
&= 1.67 \times 10 \times 10^{-25} \\
&= 1.67 \times 10^{-24}.
\end{aligned}
$$

The mass of a proton is 1.67×10^{-24} grams.

Exercise 14.4

Write in ordinary notation:

1. 4.5×10^3 2. 2.05×10^8 3. 1.903×10^{11} 4. 5.2×10^{20}

5. 9.8×10^{-1} 6. 4.02×10^{-3} 7. 1.255×10^{-6} 8. 8.806×10^{-15}

9. 3.2×10^0 10. 7.06×10^0

Write in scientific notation:

11. 25,000   12. 3,600,000

13. 500,000,000   14. 10,600,000,000

15. 2,052,000,000,000   16. 2,000,000,000,000,000,000

17. 0.00145   18. 0.000302

19. 0.000005   20. 0.00000000963

21. 0.00000000002061   22. 0.0000000000000003

Multiply, and write the result in scientific notation:

23. $(1.2 \times 10^3)(3.1 \times 10^4)$ 24. $(4.3 \times 10^9)(2.1 \times 10^6)$

25. $(5.6 \times 10^5)(4.0 \times 10^7)$

26. $(3.32 \times 10^9)(9.12 \times 10^{10})$

27. $(2.3 \times 10^{-2})(3.4 \times 10^{-5})$

28. $(3.6 \times 10^{-5})(2.0 \times 10^{-8})$

29. $(3.5 \times 10^{-5})(9.4 \times 10^{-6})$

30. $(9.23 \times 10^{-11})(8.28 \times 10^{-8})$

Solve in scientific notation to two decimal places:

31. The circumference of the earth is approximately 24,800 miles. The circumference of the sun is approximately 109 times the circumference of the earth. What is the circumference of the sun?

32. The mass of the earth is 5,970,000,000,000,000,000,000,000 kilograms. The mass of the sun is 332,000 times the mass of the earth. What is the mass of the sun?

33. The mass of a hydrogen atom is the same as the mass of a proton, 0.00000000000000000000000167 grams (Example 14.27). The mass of a uranium atom is approximately 236 times the mass of a hydrogen atom. What is the mass of a uranium atom?

34. The electric charge of an electron is 0.00000000000000000016021 coulombs. There are 92 electrons in a uranium atom. What is the total charge of the electrons in a uranium atom?

Self-test

1. Find the value:

   a. -3^4

   b. $(-3)^4$

   1a. _____

   1b. _____

2. Evaluate

   $a^2 - 3ab - b^2$

   if $a = 6$ and $b = -4$.

   2. _____

3. Simplify

   $$\frac{(rs^3t^2)^4}{(r^2s^4t)^3}$$

   3. _____

4. Write in scientific notation:

   a. 5,500,000,000

   b. 0.000000609

   4a. _____

   4b. _____

5. Multiply and write the result in scientific notation:

   $(9.3 \times 10^{-8})(2.4 \times 10^{-10})$

   5. _____

UNIT **15** Polynomials

INTRODUCTION

In this unit you will learn about a type of expression containing positive integral exponents. This type of expression is the polynomial. You will learn about some types of expressions that are polynomials, and how to add, subtract, multiply, and divide polynomials.

OBJECTIVES

When you have finished this unit you should be able to:

1. State the degree of a term and the degree of a polynomial, and write a polynomial in descending powers of a variable.
2. Find the sum or difference of two polynomials.
3. Multiply a polynomial by a monomial, and by a binomial.
4. Divide a polynomial by a monomial, and use long division to divide a polynomial by a binomial.

15.1 Definition of a Polynomial

In Section 14.2, we evaluated expressions in which the terms consisted of products of constants and variables with positive integral exponents. A **polynomial** is an expression of one or more terms, where the terms consist of such products of constants and variables. Thus, every expression in Section 14.2 is a polynomial. Some expressions in Section 14.3 are not polynomials, because we also included quotients of variables with positive integral exponents.

A **monomial** is a polynomial consisting of just one term. For example,

$$x^2y^2$$

is a monomial. Also,

$$6$$

is a monomial consisting of just a constant term.

A **binomial** is a polynomial consisting of two terms. The expression

$$x^2 + y^2$$

is a binomial.

A **trinomial** has three terms. For example,

$$a^2 - ab + b^2$$

is a trinomial. Also,

$$-x^2 + 4x + 4$$

is a trinomial that includes a constant term 4.

The **degree of a term** of a polynomial is the sum of the exponents of the variables in the term.

EXAMPLE 15.1   State the degree:

a. x^2y^2 b. $-5x^2y^2$ c. x^2y^2z

Solutions

a. There are two factors of x and two factors of y, or four factors of variables. This is the same as saying x^2 has exponent 2 and y^2 has exponent 2, and $2 + 2 = 4$. The degree of the term is 4.

b. Again there are two factors of x and two factors of y. The degree again is 4.

c. There are two factors of x, two factors of y, and one factor of z, or five factors of variables. Since $z = z^1$, we could also say that x^2 has exponent 2, y^2 has exponent 2, and z^1 has exponent 1, and $2 + 2 + 1 = 5$. The degree of the term is 5.

In part c of Example 15.1, we observed that a variable appearing without an exponent is understood to have the exponent 1. If a term has just one variable, and the variable has no exponent, the term has degree 1.

A constant term has no variables. We say that a constant term has degree 0.

EXAMPLE 15.2   State the degree:

a. $3xy$ b. $3x$ c. 3

Solutions

a. The variables x and y each are understood to have the exponent 1. Therefore, the degree of the term is 2.

b. The variable x is understood to have the exponent 1. Therefore, the degree of the term is 1.

c. The term is a constant term, in which no variable appears. Therefore, the degree of the term is 0.

The **degree of a polynomial** is the degree of its highest-degree term. To find the degree of a polynomial, we find the degree of each term, and then pick the largest of these.

EXAMPLE 15.3   State the degree of $2x^2 - 3x - 1$.

Solution

The degree of the first term is 2, the degree of the second term is 1, and the degree of the third term is 0. The largest of these numbers is 2. Therefore, the degree of the polynomial is 2.

EXAMPLE 15.4   State the degree of $3 - y^2 - 2y - y^4$.

Solution

The highest-degree term is the last term, which has degree 4. Therefore, the degree of the polynomial is 4.

EXAMPLE 15.5   State the degree of $3uv^2 - 2u^2v^2 + 3u^3$.

Solution

The degree of the first term is 3, the degree of the second term is 4, and the degree of the third term is 3. The highest-degree term has degree 4; therefore, the degree of the polynomial is 4.

When we have a polynomial in one variable, it is often convenient to write the polynomial in **descending powers** of the variable. This means that we write the highest-degree term first, then the next-highest, and so on down to the lowest-degree term.

EXAMPLE 15.6 Write $3 - y^2 - 2y - y^4$ in descending powers of y.

Solution   The highest-degree term is $-y^4$, so we write this term first. There is no term with degree 3. The next-highest-degree term is $-y^2$, so we write this term second, then $-2y$, and finally 3. In descending powers of y, the polynomial is

$$-y^4 - y^2 - 2y + 3.$$

Observe that a minus sign indicates that a term is subtracted. If no sign appears, the term is understood to be added, as in the case of the constant term 3 in the original form of the polynomial.

If a polynomial has more than one variable, it may be written in descending powers of any of its variables. If the power of the variable is the same in two terms, we write the term with the higher degree first.

EXAMPLE 15.7 Write $3uv^2 - 2u^2v^2 + 3u^3$

a. in descending powers of u.
b. in descending powers of v.

Solutions   a. The term $3u^3$ has the highest power of u, so we write this term first. The term $-2u^2v^2$ has the next highest power of u, and then $3uv^2$. In descending powers of u, the polynomial is

$$3u^3 - 2u^2v^2 + 3uv^2.$$

b. The terms $-2u^2v^2$ and $3uv^2$ have the same power of v. However, $-2u^2v^2$ has the higher degree, so we write it first and then $3uv^2$. Since v does not appear in the term $3u^3$, this term has the lowest power of v, and we write it last. In descending powers of v, the polynomial is

$$-2u^2v^2 + 3uv^2 + 3u^3.$$

Exercise 15.1

State the degree:

1. $3x^3y^3$

2. $5x^2y$

3. $-2xyz^2$

4. $-xyz$

5. 2

6. $2x$

7. $x^4 - x^3 - x^2 + x$

8. $y^5 + y^4 - y - 1$

9. $2x^3 - 3x^6 + 12$

10. $4z + 3z^2 - 2z^3 - 5z^4$

11. $2x^2y^2 - 3x^2y^3 + 4x^2y^4$

12. $u^3 + u^2v^2 + uv^4$

Write in descending powers of x:

13. $x - x^3 + x^4 - x^2$

14. $3 + 2x - x^2$

15. $2x^2 - 4 + x$

16. $4 + x^4 - 3x - 4x^3$

17. $y^3 - 3xy^2 + 3x^2y - x^3$

18. $xy + x^3 - x^2y^2$

Write in descending powers of y:

19. $x^4 - 4x^3y + 6x^2y^2 - 4xy^3 + y^4$

20. $xy + x^3 - x^2y^2$

15.2   Addition and Subtraction

Addition and subtraction of polynomials is exactly the same as addition and subtraction of linear algebraic expressions. We remove the parentheses, and combine any resulting like terms. Recall that like terms must have exactly the same literal parts. For example, $5x^2$ and $-2x^2y$ are not like terms; however, $5x^2y$ and $-2x^2y$ are like terms.

EXAMPLE 15.8   Add $(3x^2 - 2x - 1) + (-x^2 + 4x - 5)$.

Solution   First, we remove the parentheses. Since the operation is addition, all the signs stay as they are:

$$(3x^2 - 2x - 1) + (-x^2 + 4x - 5) = 3x^2 - 2x - 1 - x^2 + 4x - 5.$$

Now, we combine the like terms. The x^2 terms are like terms, the x terms are like terms, and the constant terms are like terms:

$$(3x^2 - 2x - 1) + (-x^2 + 4x - 5) = 3x^2 - 2x - 1 - x^2 + 4x - 5$$
$$= 2x^2 + 2x - 6.$$

There are no further like terms, and so the addition is complete.

EXAMPLE 15.9   Add $(5 + z^4 - z^2) + (2z^3 - z^2 - 6)$.

Solution   It is not necessary to write the polynomials in descending powers of z. We just remove the parentheses and combine the like terms:

$$(5 + z^4 - z^2) + (2z^3 - z^2 - 6) = 5 + z^4 - z^2 + 2z^3 - z^2 - 6$$
$$= z^4 + 2z^3 - 2z^2 - 1.$$

It is usual to write the sum in descending powers of the variable.

EXAMPLE 15.10   Add $(u^3 + u^2v) + (v^3 + uv^2)$.

Solution   Removing the parentheses, we have

$$(u^3 + u^2v) + (v^3 + uv^2) = u^3 + u^2v + v^3 + uv^2$$
$$= u^3 + u^2v + uv^2 + v^3.$$

Observe that there are no like terms; therefore, the addition is complete. We have written the sum in descending powers of u.

To subtract polynomials, we proceed as in subtracting linear expressions. We remove the parentheses and use the opposite sign for each term of the second polynomial. Then, we combine like terms as before.

EXAMPLE 15.11   Subtract $(2x^2 - 5x + 6) - (x^2 - 3x + 2)$.

Solution   First, we remove the parentheses, using the opposite sign for each term of the second polynomial:

$$(2x^2 - 5x + 6) - (x^2 - 3x + 2) = 2x^2 - 5x + 6 - x^2 + 3x - 2.$$

Now, we combine the like terms:

$$(2x^2 - 5x + 6) - (x^2 - 3x + 2) = 2x^2 - 5x + 6 - x^2 + 3x - 2$$
$$= x^2 - 2x + 4.$$

EXAMPLE 15.12   Subtract $(5 - x^3 + 2x^2) - (10 - x^3 - 3x)$.

Solution   We remove the parentheses and combine like terms:

$$(5 - x^3 + 2x^2) - (10 - x^3 - 3x) = 5 - x^3 + 2x^2 - 10 + x^3 + 3x$$
$$= 2x^2 + 3x - 5.$$

Observe that $-x^3 + x^3 = 0 \cdot x^3$; therefore, no x^3 term appears in the resulting polynomial.

EXAMPLE 15.13   Subtract $(x^2y - x^2y^2) - (y^2x - 2y^2x^2)$.

Solution   We remove the parentheses and combine like terms, observing that $2y^2x^2 = 2x^2y^2$:

$$(x^2y - x^2y^2) - (y^2x - 2y^2x^2) = x^2y - x^2y^2 - y^2x + 2y^2x^2$$
$$= x^2y - x^2y^2 - y^2x + 2x^2y^2$$
$$= x^2y + x^2y^2 - y^2x$$
$$= x^2y^2 + x^2y - xy^2,$$

where we have written the resulting polynomial in descending powers of x.

Exercise 15.2

Add or subtract as indicated:

1. $(x^2 - 2x - 3) + (x^2 - 3x + 2)$

2. $(x^2 - 5x + 1) + (-2x^2 + x + 2)$

3. $(6 - 3x^2 - x) + (2x + 4x^2 - 1)$

4. $(y + 4y^3 + 8y^2) + (3y - y^3 - 9y^2)$

5. $(2x^2 - 3x + 1) - (x^2 - 5x - 6)$

6. $(5y^2 + 8y - 4) - (-3y^2 - 2y + 7)$

7. $(z - 9 - 4z^2) - (6 - z - 2z^2)$

8. $(4u^2 - 2u - 7) - (5 - 2u - 3u^2)$

9. $(3 - x^4 + 2x^3) + (2x^2 + 3x^4 - 6)$

10. $(y^4 - 4y^2 - 3) - (y^4 - 6y^3 + 2y)$

11. $(y^3 - y^2 + 4) - (y^2 - y + 4)$

12. $(3z^2 - 4z + 6) + (7 - 6z - 3z^2)$

13. $(u^2 + uv) + (vu + v^2)$

14. $(x^3 + x^2y) + (xy^2 + y^3)$

15. $(x^2y^2 - xy^2) - (yx^2 - y^2x^2)$

16. $(s^3 - 3s^2t + 2st^2) - (3st^2 + 2s^2t - t^3)$

15.3 Multiplication

We begin by multiplying two monomials. Recall from Section 14.3 that to multiply powers with the same base we add the exponents.

EXAMPLE 15.14 Multiply $(3x^2y)(4x^3z)$.

Solution   We multiply the numerical coefficients. Then, to multiply x^2 and x^3 we add the exponents to obtain x^5. Finally, writing all the factors, with the numerical coefficient first and the variables in alphabetical order.

$$(3x^2y)(4x^3z) = 12x^5yz.$$

EXAMPLE 15.15 Multiply $(5pq^2)(-2p^2q)$.

Solution   We multiply 5 and -2 to obtain -10. Then, multiplying the factors containing variables by adding the exponents, we have

$$(5pq^2)(-2p^2q) = -10p^3q^3.$$

Multiplication of a polynomial by a monomial is done in the same way as multiplication by a constant. We use the distributive property and multiply each term of the polynomial by the monomial.

EXAMPLE 15.16 Multiply $4(3x^2 + 2y^2)$.

Solution   In this example the monomial is a constant. We multiply each term of the polynomial by 4:

$$4(3x^2 + 2y^2) = 4(3x^2) + 4(2y^2)$$
$$= 12x^2 + 8y^2.$$

EXAMPLE 15.17 Multiply $-3x^2(2x^2 - 3x + 5)$.

Solution   We multiply each term of the polynomial by the monomial $-3x^2$:

$$-3x^2(2x^2 - 3x + 5) = (-3x^2)(2x^2) - (-3x^2)(3x) + (-3x^2)(5).$$

Now, we multiply each term, multiplying the powers by adding the exponents:

$$-3x^2(2x^2 - 3x + 5) = (-3x^2)(2x^2) - (-3x^2)(3x) + (-3x^2)(5)$$
$$= -6x^4 + 9x^3 - 15x^2.$$

EXAMPLE 15.18 Multiply $6x^3y^3(x^2 - 2xy + 2y^2)$.

Solution   We multiply each term of the polynomial by the monomial $6x^3y^3$. Then, we multiply each term, multiplying powers of the same variable by adding their exponents:

$$6x^3y^3(x^2 - 2xy + 2y^2) = (6x^3y^3)(x^2) - (6x^3y^3)(2xy) + (6x^3y^3)(2y^2)$$
$$= 6x^5y^3 - 12x^4y^4 + 12x^3y^5.$$

Using the distributive property twice, we can derive a method for multiplying two binomials. Suppose, for example, we wish to multiply $2x - 3$ by $x + 2$. We use A to represent the second binomial $x + 2$:

$$(2x - 3)(x + 2) = (2x - 3)A$$
$$= (2x)A - (3)A.$$

Now we substitute $x + 2$ back for A, and use the distributive property a second time:

$$(2x)A - (3)A = 2x(x + 2) - 3(x + 2)$$
$$= 2x^2 + 4x - 3x - 6$$
$$= 2x^2 + x - 6.$$

Thus,

$$(2x - 3)(x + 2) = 2x^2 + x - 6.$$

Leaving out the use of A, we can write

$$(2x - 3)(x + 2) = 2x(x + 2) - 3(x + 2)$$
$$= 2x^2 + 4x - 3x - 6$$
$$= 2x^2 + x - 6.$$

Observe that the expression derived in the middle step,

$$(2x - 3)(x + 2) = 2x^2 + 4x - 3x - 6,$$

is the result of multiplying

the two first terms, $2x$ and x,
the two outside terms, $2x$ and 2,
the two inside terms, -3 and x, and
the two last terms, -3 and 2.

This observation leads to a shorter method for multiplying two binomials. Mentally, we use the diagram

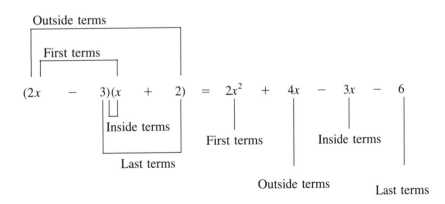

Then, combining the results of the outside and inside terms,

$$(2x - 3)(x + 2) = 2x^2 + 4x - 3x - 6$$
$$= 2x^2 + x - 6.$$

We call the combined result of the outside and inside terms the **middle term.** In this example, the middle term is x.

EXAMPLE 15.19   Multiply $(x + 2)(x - 4)$.

Solution   Using the diagram, the products are

first terms: $(x)(x) = x^2$,
outside terms: $(x)(-4) = -4x$,
inside terms: $(2)(x) = 2x$,
last terms: $(2)(-4) = -8$.

Therefore,

$$(x + 2)(x - 4) = x^2 - 4x + 2x - 8$$
$$= x^2 - 2x - 8.$$

EXAMPLE 15.20   Multiply $(5x - 3)(2x - 4)$.

Solution   Calculating, in order, the products of first terms, outside terms, inside terms, and last terms, we have

$$(5x - 3)(2x - 4) = 10x^2 - 20x - 6x + 12$$
$$= 10x^2 - 26x + 12.$$

EXAMPLE 15.21   Multiply $(x^2 - 3)(x^2 + 2)$.

Solution   The product of the first terms is $(x^2)(x^2) = x^4$, and the products of the outside and the inside terms are $(x^2)(2) = 2x^2$ and $(-3)(x^2) = -3x^2$:

$$(x^2 - 3)(x^2 + 2) = x^4 + 2x^2 - 3x^2 - 6$$
$$= x^4 - x^2 - 6.$$

There are two special cases of multiplying two binomials. One case is the **perfect square.**

EXAMPLE 15.22   Multiply $(2x + 3)^2$.

Solution $(2x + 3)^2$ means $(2x + 3)(2x + 3)$:

$$(2x + 3)(2x + 3) = 4x^2 + 6x + 6x + 9$$
$$= 4x^2 + 12x + 9.$$

It is a common error to think that $(2x + 3)^2$ is the same as $(2x)^2 + 3^2$, or $4x^2 + 9$. Observe that we also have a *middle term* $12x$, which is the combined products of the inside and outside terms, or

$$(2x)(3) + (3)(2x) = 6x + 6x.$$

The second special case of multiplying two binomials is the **sum and difference.**

☐ **EXAMPLE 15.23** Multiply $(x + 2y)(x - 2y)$.

Solution Observe that one binomial is the sum and the other is the difference of the same two terms, x and $2y$:

$$(x + 2y)(x - 2y) = x^2 - 2xy + 2xy - 4y^2$$
$$= x^2 - 4y^2.$$

Since $-2xy + 2xy = 0$, no middle term appears. You should be careful not to confuse the sum and difference with the perfect square. In a sum and difference, a middle term *never* appears. In a perfect square, a middle term *always* appears.

If we have a product of a binomial and a higher order polynomial, we extend the distributive property one more step. The result is that we multiply every term of the polynomial by the first term of the binomial, and also every term of the polynomial by the second term of the binomial.

☐ **EXAMPLE 15.24** Multiply $(2x + 3)(3x^2 - 2x + 4)$.

Solution We multiply each term of the polynomial by $2x$ and also by 3:

$$(2x + 3)(3x^2 - 2x + 4) = (2x)(3x^2) - (2x)(2x) + (2x)(4) + (3)(3x^2) - (3)(2x) + (3)(4)$$
$$= 6x^3 - 4x^2 + 8x + 9x^2 - 6x + 12$$
$$= 6x^3 + 5x^2 + 2x + 12.$$

Often it is convenient to write the first polynomial beneath the second, and then multiply each term of the top polynomial by each term of the bottom polynomial.

☐ **EXAMPLE 15.25** Multiply $(x - 3y)(x^2 + 3xy + 9y^2)$.

Solution We write $x - 3y$ beneath $x^2 + 3xy + 9y^2$:

$$\begin{array}{r} x^2 + 3xy + 9y^2 \\ x - 3y \\ \hline x^3 + 3x^2y + 9xy^2 \\ - 3x^2y - 9xy^2 - 27y^3 \\ \hline x^3 \qquad\qquad - 27y^3. \end{array}$$

In the first line, we have multiplied each term of $x^2 + 3xy + 9y^2$ by x. In the second line, we have multiplied each term of $x^2 + 3xy + 9y^2$ by $-3y$. For convenience, we have written like terms beneath one another. Then, we have added downward to obtain the product:

$$(x - 3y)(x^2 + 3xy + 9y^2) = x^3 - 27y^3.$$

☐ **Exercise 15.3**

Multiply:

1. $(4x^2y)(5yz^2)$

2. $(6ab)(6a^2c^2)$

3. $(5st)(-3s^2t^2)$

4. $(3p^3qr)(-2pq^2r)$

5. $6(3x^2 - 2)$

6. $-5(2x^2 - 3y^2)$

7. $-2x(4x^2 - x + 3)$

8. $3x^3(x^4 - 4x^2 + 4)$

9. $2xy(x^2y^2 - 2x^2y + xy)$

10. $x^2y^3(x^3 - 3x^2y - 3y^2)$

11. $(x + 1)(x + 2)$

12. $(x - 2)(x + 3)$

13. $(x - 4)(3x + 2)$

14. $(4x - 5)(x - 6)$

15. $(3x + 4)(2x - 1)$

16. $(2x - 1)(5x - 8)$

17. $(x^2 - 3)(x^2 - 5)$

18. $(x^2 + 3)(2x^2 - 7)$

19. $(3x - 2y)(x + 5y)$

20. $(4x - 3y)(3x - 4y)$

21. $(3x + 5)(3x + 5)$

22. $(2x - 1)(2x - 1)$

23. $(x - 5)^2$

24. $(4x + 3)^2$

25. $(4x - 3)(4x + 3)$

26. $(6x + 5)(6x - 5)$

27. $(x + 2)(x^2 - 4x + 4)$

28. $(2x - 3)(2x^2 - x + 3)$

29. $(x - y)(x^2 + xy + y^2)$

30. $(3x - 2y)(2x^2 - 3xy - 3y^2)$

31. $(x^2 - 2x + 1)(x^2 - 4x + 4)$

32. $(x + y + 2)(x - y - 2)$

15.4 Division

To divide a polynomial by a monomial, we divide each term of the polynomial by the monomial.

EXAMPLE 15.26   Divide $\dfrac{2x + 4y}{2}$.

Solution   We divide each term of the polynomial $2x + 4y$ by 2:

$$\frac{2x + 4y}{2} = \frac{2x}{2} + \frac{4y}{2}$$
$$= x + 2y.$$

EXAMPLE 15.27   Divide $\dfrac{6x + y}{3}$.

Solution   The second term of $6x + y$ is not divisible by 3. We write

$$\frac{6x + y}{3} = \frac{6x}{3} + \frac{y}{3}$$
$$= 2x + \frac{y}{3}.$$

It is a common error to divide only the term $6x$ by 3. We must divide *each term* by 3. Since the numerator of the second term is not divisible by 3, we leave the second term as a fraction.

☐ EXAMPLE 15.28   Divide $\dfrac{16x^2y^3 - 4xy^2 + y}{4xy}$.

Solution   We divide each term of the polynomial by the monomial:

$$\frac{16x^2y^3 - 4xy^2 + y}{4xy} = \frac{16x^2y^3}{4xy} - \frac{4xy^2}{4xy} + \frac{y}{4xy}.$$

Then, we recall from Section 14.3 that to divide powers with the same base we subtract the exponents:

$$\frac{16x^2y^3 - 4xy^2 + y}{4xy} = \frac{16x^2y^3}{4xy} - \frac{4xy^2}{4xy} + \frac{y}{4xy}$$

$$= 4xy^2 - y + \frac{1}{4x}.$$

Before we discuss the division of a polynomial by a binomial, we review long division of whole numbers. Consider the division

$$36\overline{)774}$$

As a first step, we simply consider 7 divided by 3. The nearest partial quotient is 2:

$$\begin{array}{r} 2 \\ 36\overline{)774} \end{array}$$

Now, we multiply 36 by 2:

$$\begin{array}{r} 2 \\ 36\overline{)774} \\ 72 \end{array}$$

Then, we subtract 72 from 77:

$$\begin{array}{r} 2 \\ 36\overline{)774} \\ \underline{72} \\ 5 \end{array}$$

We bring down the 4 and repeat the process:

$$\begin{array}{r} 21 \\ 36\overline{)774} \\ \underline{72} \\ 54 \\ \underline{36} \\ 18 \end{array}$$

Since there are no more numbers to bring down, the division is complete. We have a remainder of 18. We may write the remainder as a fraction in which the denominator is the divisor. Therefore, the quotient is

$$21\frac{18}{36} = 21\frac{1}{2}.$$

To divide a polynomial by a binomial, we use essentially the same process.

EXAMPLE 15.29   Divide $\dfrac{x^2 + 5x + 6}{x + 2}$.

Solution   We write

$$x + 2 \overline{)x^2 + 5x + 6}.$$

As a first step, we consider x^2 divided by x. The quotient is x. Therefore, we write an x above the x term of the polynomial being divided:

$$\begin{array}{r} x \\ x + 2 \overline{)x^2 + 5x + 6} \end{array}$$

Now, we multiply $x + 2$ by x, being careful to multiply both terms:

$$\begin{array}{r} x \\ x + 2 \overline{)x^2 + 5x + 6} \\ x^2 + 2x \end{array}$$

Then, we subtract $x^2 + 2x$ from $x^2 + 5x$:

$$(x^2 + 5x) - (x^2 + 2x) = x^2 + 5x - x^2 - 2x$$
$$= 3x.$$

If our division is correct, the x^2 term should not appear in the remainder:

$$\begin{array}{r} x \\ x + 2 \overline{)x^2 + 5x + 6} \\ x^2 + 2x \\ \hline 3x \end{array}$$

We bring down the 6, divide $3x$ by x to obtain 3, and multiply $x + 2$ by 3:

$$\begin{array}{r} x + 3 \\ x + 2 \overline{)x^2 + 5x + 6} \\ x^2 + 2x \\ \hline 3x + 6 \\ 3x + 6 \end{array}$$

Since $(3x + 6) - (3x + 6) = 0$, the remainder is 0, and the quotient is

$$x + 3.$$

EXAMPLE 15.30   Divide $\dfrac{x^2 - 4x - 5}{x - 3}$.

Solution   We write

$$x - 3 \overline{)x^2 - 4x - 5}$$

and divide x^2 by x:

$$x - 3 \overline{)\, x^2 - 4x - 5 \,}^{\;x}$$

Multiplying $x - 3$ by x,

$$x - 3 \overline{)\, x^2 - 4x - 5 \,}^{\;x}$$
$$x^2 - 3x$$

Subtracting, we have

$$(x^2 - 4x) - (x^2 - 3x) = x^2 - 4x - x^2 + 3x$$
$$= -x.$$

Observe that we may indicate subtraction by opposite signs for the second polynomial:

$$x - 3 \overline{)\, x^2 \ominus 4x - 5 \,}^{\;x}$$
$$\ominus\; x^2 \oplus 3x$$
$$\overline{\; -\; x}$$

We bring down -5 and divide $-x$ by x to obtain -1. Then, multiplying $x - 3$ by -1,

$$x - 3 \overline{)\, x^2 \ominus 4x - 5 \,}^{\;x - 1}$$
$$\ominus\; x^2 \oplus 3x$$
$$\overline{\; -\; x\; -\; 5}$$
$$\; -\; x\; +\; 3$$

Subtracting, we have

$$(-x - 5) - (-x + 3) = -x - 5 + x - 3$$
$$= -8,$$

or

$$x - 3 \overline{)\, x^2 - 4x - 5 \,}^{\;x - 1}$$
$$\ominus\; x^2 \oplus 3x$$
$$\overline{\; -\; x\; -\; 5}$$
$$\oplus \ominus$$
$$\overline{\; -\; x\; +\; 3}$$
$$\overline{\; -\; 8}$$

The remainder is -8. We write this as the fraction $\dfrac{-8}{x - 3}$. The quotient is

$$x - 1 + \frac{-8}{x - 3}.$$

☐ **EXAMPLE 15.31**   Divide $\dfrac{y^3 - 5y^2 + y + 10}{y - 4}$.

Solution   We follow the same steps as in the preceding examples: divide y^3 by y to obtain y^2; multiply back by y^2; subtract, remembering to indicate the opposite signs; and repeat the process, bringing down y and then 10:

$$
\begin{array}{r}
y^2 - y - 3 \\
y - 4 \overline{)\, y^3 \ominus 5y^2 + y + 10} \\
\ominus\ y^3 \oplus 4y^2 \\
\hline
- y^2 + y \\
\oplus\ \ominus \\
y^2 \mp 4y \\
\hline
- 3y + 10 \\
\oplus\ \ominus \\
3y \mp 12 \\
\hline
-2
\end{array}
$$

The quotient is

$$
y^2 - y - 3 + \frac{-2}{y - 4}.
$$

If the terms in a polynomial are not in descending powers of the variable, we must write them in descending powers before dividing.

EXAMPLE 15.32   Divide $\dfrac{4z^3 - z + 2z^2 - 1}{z - 1}$.

Solution   We write the polynomial in descending powers of z and then divide as before:

$$
\begin{array}{r}
4z^2 + 6z + 5 \\
z - 1 \overline{)\, 4z^3 + 2z^2 - z - 1} \\
\ominus\ 4z^3 \oplus 4z^2 \\
\hline
6z^2 \ominus z \\
\ominus\ 6z^2 \oplus 6z \\
\hline
5z \ominus 1 \\
\ominus\ 5z \oplus 5 \\
\hline
4
\end{array}
$$

The quotient is

$$
4z^2 + 6z + 5 + \frac{4}{z - 1}.
$$

If one or more powers of a variable do not appear, we must allow for these powers in the division process. A convenient way is to write in 0 times the power.

EXAMPLE 15.33   Divide $\dfrac{x^3 - 4x + 2}{x + 2}$.

Solution   We write $0x^2$ to allow for the x^2 term:

$$
\begin{array}{r}
x^2 \;-\; 2x \;+\; 0 \\[2pt]
x + 2 \overline{)\, x^3 \;+\; 0x^2 \;-\; 4x \;+\; 2 \,} \\[2pt]
\ominus \; x^3 \;+\; 2x^2 \\[2pt]
\hline
2x^2 \;-\; 4x \\[2pt]
\oplus \; 2x^2 \;+\; 4x \\[2pt]
\hline
0x \;+\; 2 \\[2pt]
\ominus \; 0x \;+\; 0 \\[2pt]
\hline
2
\end{array}
$$

The quotient is

$$x^2 \;-\; 2x \;+\; \frac{2}{x + 2}.$$

Exercise 15.4

Divide:

1. $\dfrac{5x + 10y}{5}$

2. $\dfrac{6x - 15y}{3}$

3. $\dfrac{2x - 3}{2}$

4. $\dfrac{8x^2 + 10x}{4x}$

5. $\dfrac{6x^2y^2 + 12xy + 6}{6x}$

6. $\dfrac{3x^2y - 6xy^2 + 9y^3}{3xy}$

7. $\dfrac{x^2 + 5x + 6}{x + 3}$

8. $\dfrac{x^2 - x - 2}{x + 1}$

9. $\dfrac{x^2 - 7x + 8}{x - 4}$

10. $\dfrac{2x^2 - 5x + 5}{x - 2}$

11. $\dfrac{x^3 - 3x^2 + 3x - 1}{x - 1}$

12. $\dfrac{4x^3 - 2x^2 + 6x - 2}{2x - 1}$

13. $\dfrac{3 - 2y^2 - 5y + y^3}{y - 3}$

14. $\dfrac{4z^3 - 2z - 6z^2 - 3}{2z - 3}$

15. $\dfrac{y^2 + y^3 - 3}{y - 2}$

16. $\dfrac{x^3 - 1}{x - 1}$

Self-test

1. State the degree:

   a. $-5s^3t$

   b. $2x^3 + 3x^2 - 4x^4 - 4$

   1a. _____

   1b. _____

Multiply:

2. $(2x - 3)(4x + 5)$

   2. _____

3. $(x - 3y)(x^2 - xy + 2y^2)$

   3. _____

4. Subtract:

   $(4x^2 - 6y^2 + 2xy) - (3xy - 2x^2 - 3y^2)$

   4. _____

5. Divide:

   $$\frac{y + 2y^3 - 4}{y - 2}$$

   5. _____

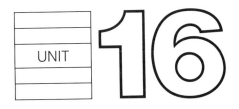

UNIT

16 Cumulative Review

INTRODUCTION
In this unit you should make certain you remember all the material in all of the preceding units.

OBJECTIVE
When you have finished this unit you should be able to demonstrate that you can fulfill every objective of each preceding unit.

To prepare for this unit you should review the Self-Tests for Units 1 through 15. Do each problem of each Self-Test over again. If you cannot do a problem, or even if you have the slightest difficulty, you should:

1. Find out from the answer section the Objective for the unit to which the problem relates.
2. Review all the material in the section which has the same number as the objective, and redo all the Exercises for the section.
3. Try the Self-Test for the unit again.

Repeat these steps until you can do each problem in each Self-Test for each of Units 1 through 15 easily and accurately.

Self-test

1. Evaluate:

   a. $-3 - (-9)$

   b. $-3(-9)$

   1a. _____

   1b. _____

Solve and check:

2. $(2x - 3) - (4x - 6) = 2 - (x - 4)$

   2. _____

3. $3 - 9x > 5 - 3x$

   3. _____

4. $3x + 2y = 4$
   $x - 3y = 5$

   4. _____

5. 545 tickets to a show were sold for a total of $1242.50. Orchestra seats cost $2.50 and balcony seats cost $1.50. How many tickets for orchestra seats and how many tickets for balcony seats were sold?

   5. _____

6. Find the *y*-intercept, *x*-intercept, slope, and draw the graph of $2x + 4y - 6 = 0$.

y-intercept _____

x-intercept _____

slope _____

7. Draw the graph of $y - 2x + 4 < 0$.

8. Evaluate $-t^2 - 3t + 4$ for $t = -2$.

8. _____

9. Simplify $\left(\dfrac{a^4 b}{a^2 b^3}\right)^5$.

9. _____

10. Multiply:

a. $(x - 3y)^2$

b. $(3x - 4y)(3x + 4y)$

10a. _____

10b. _____

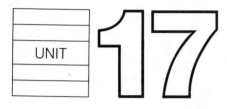

Factoring

INTRODUCTION

In dealing with expressions that are not linear, it is often helpful to be able to rewrite an expression as a product of expressions that are linear. When an expression cannot be rewritten as linear expressions, it may be possible to write a product of expressions that are of lower degree. In this unit you will learn how to factor an expression into two or more expressions of lower degree. First, you will learn how to factor out a common a factor. Then, you will learn some special cases, and finally, how to factor general trinomials.

OBJECTIVES

When you have finished this unit you should be able to:

1. Factor out a common factor from a polynomial.
2. Factor the difference of two squares as a sum and difference of binomials.
3. Factor a perfect trinomial square as the square of a binomial.
4. Factor a trinomial into two binomial factors.

17.1 Common Factors

In other courses you have written positive integers as products of two or more positive integers. The integers in the products are called **factors.** For example, we say 2 and 6 are factors of 12 since $12 = 2 \cdot 6$. Also, 3 and 4 are factors of 12 since $12 = 3 \cdot 4$. If we continue to find positive integral factors, we have

$$12 = 2 \cdot 6 = 2 \cdot 2 \cdot 3$$

and

$$12 = 3 \cdot 4 = 3 \cdot 2 \cdot 2.$$

When no further factorization is possible, the factorization is **complete.** The final factors in any complete factorization are the same. We say the factorization of an integer is **unique.** Observe that, to obtain a unique factorization, we use only integers. We do not use, for example, fractions such as $12 = \frac{1}{2} \cdot 24$. When we use only integers, we say that the factorization is **over the integers.**

Similarly, some polynomials can be factored over the integers. When such a factorization exists, the complete factorization is unique. Recall from previous units that we used the distributive property to write, for example,

$$2x + 2y = 2(x + y).$$

The expression $2(x + y)$ is the unique factorization of the polynomial $2x + 2y$ over the integers. The factor 2 appears in each term of the polynomial. We say that 2 is a **common factor.** The simplest type of factoring is factoring out a common factor.

EXAMPLE 17.1   Factor $6x + 12y$.

Solution   There are several common factors, in particular, 2, 3, and 6. However, if we factor out 2 or 3, we have $2(3x + 6y)$ or $3(2x + 4y)$. These factorizations are not complete. We choose the largest common factor, 6. Then,

$$6x + 12y = 6(x + 2y).$$

Whenever we say to factor a polynomial, we mean the *complete* factorization over the integers.

EXAMPLE 17.2   Factor $4x^2 + 12x$.

Solution   The largest integral common factor is 4:

$$4x^2 + 12x = 4(x^2 + 3x).$$

However, there is a variable common factor x, so the factorization is not yet complete. We factor out the common factor x, recalling that $x^2 = x \cdot x$:

$$4x^2 + 12x = 4(x^2 + 3x)$$
$$= 4x(x + 3).$$

EXAMPLE 17.3   Factor $5x^2y + 5xy^2$.

Solution   We look for the largest common factor, including integers and variables. There are common factors of 5, x, and y. Therefore, the largest common factor is $5xy$:

$$5x^2y + 5xy^2 = (5xy)(x) + (5xy)(y)$$
$$= 5xy(x + y).$$

To check this factorization, we first check that the resulting polynomial has no further common factors. Then, we check the factorization by multiplying:

$$5xy(x + y) = 5xy(x) + 5xy(y)$$
$$= 5x^2y + 5xy^2.$$

EXAMPLE 17.4   Factor $2xy - y$.

Solution   The only common factor is y. Observe that $y = (y)1$, therefore

$$2xy - y = (y)(2x) - (y)(1)$$
$$= y(2x - 1).$$

To check, we observe that $2x - 1$ has no common factors, and also

$$y(2x - 1) = y(2x) - y(1)$$
$$= 2xy - y.$$

If the polynomial has three or more terms, we look for factors common to all the terms.

EXAMPLE 17.5   Factor $4x^3 - 6x^2 + 6x$.

Solution   The factors 2 and x are common to all terms:

$$4x^3 - 6x^2 + 6x = (2x)(2x^2) - (2x)(3x) + (2x)(3)$$
$$= 2x(2x^2 - 3x + 3).$$

You should observe that the resulting polynomial has no further common factors and check by multiplying.

Exercise 17.1

Factor:

1. $3x + 3y$

2. $2x - 4$

3. $6x + 3$

4. $2x^2 - 6x + 2$

5. $x^2 - 2x$

6. $2x^2 + 4x$

7. $2xy - 2x^2$

8. $4x^2y^2 + 16xy$

9. $x^3y - x^2y$

10. $xy^2 - 3xy^3$

11. $10x^2 - 10x - 5$

12. $12x^2 + 8x - 16$

13. $x^3 + 2x^2 + 2x$

14. $3x^3 - 6x^2 - 3x$

15. $x^3y - x^2y^2 + xy^3$

16. $4x^2y^3 + 8x^2y^2 + 12x^2y$

17.2   Difference of Squares

Recall from Section 15.3 the special case of multiplying two binomials that we called the sum and difference. In Example 15.23, we multiplied

$$(x + 2y)(x - 2y) = x^2 - 2xy + 2xy - 4y^2$$
$$= x^2 - 4y^2.$$

Observe that x^2 is a square, and also that $4y^2 = (2y)^2$ is a square. We call the result a **difference of squares.** A difference of squares is a binomial, where each term is a square, and the second term is subtracted from the first.

Whenever we have a difference of squares, we can factor the polynomial into a sum and difference. That is, a binomial of the form $a^2 - b^2$ can be factored into the sum and difference

$$a^2 - b^2 = (a + b)(a - b).$$

EXAMPLE 17.6   Factor $x^2 - 9$.

Solution x^2 is a square, and 9 is the square 3^2. Therefore, we have a difference of squares:

$$x^2 - 9 = x^2 - 3^2$$
$$= (x + 3)(x - 3).$$

To check, we multiply the resulting binomials:

$$(x + 3)(x - 3) = x^2 - 3x + 3x - 9$$
$$= x^2 - 9.$$

EXAMPLE 17.7   Factor $x^2 - y^2$.

Solution   We have the difference of the squares x^2 and y^2. Therefore,

$$x^2 - y^2 = (x + y)(x - y).$$

To check, we multiply

$$(x + y)(x - y) = x^2 - xy + xy - y^2$$
$$= x^2 - y^2.$$

EXAMPLE 17.8   Factor $64x^2 - 1$.

Solution $64x^2$ is the square $(8x)^2$. Also, $1 = 1^2$, and so we have a difference of the squares $(8x)^2$ and 1^2. Therefore,

$$64x^2 - 1 = (8x)^2 - 1^2$$
$$= (8x + 1)(8x - 1).$$

You should check this factorization by multiplying.

We may also use the difference of squares form to factor polynomials that are differences of squares of squares, squares of cubes, and so on.

EXAMPLE 17.9   Factor $9x^4 - 16y^4$.

Solution   We observe that $9x^4 = (3x^2)^2$ and $16y^4 = (4y^2)^2$. Therefore, we have a difference of squares:

$$9x^4 - 16y^4 = (3x^2)^2 - (4y^2)^2$$
$$= (3x^2 + 4y^2)(3x^2 - 4y^2).$$

To check, we multiply:

$$(3x^2 + 4y^2)(3x^2 - 4y^2)$$
$$= (3x^2)(3x^2) - (3x^2)(4y^2) + (3x^2)(4y^2) - (4y^2)(4y^2)$$
$$= 9x^4 - 12x^2y^2 + 12x^2y^2 - 16y^4$$
$$= 9x^4 - 16y^4.$$

We always factor out any common factors before using the difference of squares form.

EXAMPLE 17.10   Factor $4x^4 - 16x^2$.

Solution   There is a common factor $4x^2$. We factor out the common factor and then factor the resulting difference of squares:

$$4x^4 - 16x^2 = 4x^2(x^2 - 4)$$
$$= 4x^2(x + 2)(x - 2).$$

To check, you should first multiply the sum and difference, and then multiply by the monomial.

☐☐☐☐ Exercise 17.2

Factor:

1. $x^2 - 1$ 2. $x^2 - 36$ 3. $4x^2 - 9$ 4. $25x^2 - 64$

5. $x^2 - 9y^2$ 6. $49x^2 - 16y^2$ 7. $16x^2 - 1$ 8. $4x^2y^2 - 1$

9. $64x^2 - y^2$ 10. $x^2 - 144y^2$ 11. $x^4 - 4$ 12. $49x^4 - 1$

13. $4x^4 - 25y^4$ 14. $36x^6 - y^6$ 15. $16x^2 - 4$ 16. $4x^2y^2 - y^2$

17. $100x^4 - 25y^4$ 18. $36x^4 - 64x^2$ 19. $x^8 - 36x^2$ 20. $16x^8 - 64y^4$

☐ 17.3 ☐ Perfect Squares

In Section 15.3 we saw a special case of multiplying two binomials, which we called the perfect square. In Example 15.22 we multiplied out a perfect square:

$$(2x + 3)^2 = (2x + 3)(2x + 3)$$
$$= 4x^2 + 6x + 6x + 9$$
$$= 4x^2 + 12x + 9.$$

We observed that $4x^2$ is the square $(2x)^2$ and 9 is the square 3^2. But we also have a middle term $12x = 2(6x) = 2(2x)(3)$.

A trinomial can be factored into a perfect square when the first and last terms are squares, and the middle term is twice the product of the quantities squared. That is, a trinomial of the form $a^2 + 2ab + b^2$ can be factored into the perfect square

$$a^2 + 2ab + b^2 = (a + b)^2.$$

Also, since $b^2 = (-b)^2$,

$$a^2 - 2ab + b^2 = (a - b)^2.$$

☐☐☐☐ EXAMPLE 17.11   Factor $x^2 + 6x + 9$.

Solution x^2 is a square, and 9 is the square 3^2. Also, the middle term is $6x = 2(x)(3)$. Therefore, we have a perfect square:

$$x^2 + 6x + 9 = x^2 + 2(x)(3) + 3^2$$
$$= (x + 3)^2.$$

To check, we multiply:

$$(x + 3)(x + 3) = x^2 + 3x + 3x + 9$$
$$= x^2 + 6x + 9.$$

☐☐☐☐ EXAMPLE 17.12   Factor $x^2 - 10x + 25$.

Solution x^2 is a square, and 25 is either the square 5^2, or the square $(-5)^2$. Also, the middle term is $-10x = 2(x)(-5)$. Therefore, we have a perfect square:

$$x^2 - 10x + 25 = x^2 + 2(x)(-5) + (-5)^2$$
$$= (x - 5)^2.$$

You should check this perfect square by multiplying.

EXAMPLE 17.13   Factor $4x^2 + 20x + 25$.

Solution $4x^2$ is the square $(2x)^2$ and 25 is the square 5^2. The middle term is $20x = 2(2x)(5)$. Therefore, we have the perfect square

$$4x^2 + 20x + 25 = (2x)^2 + 2(2x)(5) + 5^2$$
$$= (2x + 5)^2.$$

You should check this perfect square by multiplying.

EXAMPLE 17.14   Factor $x^2 - 2xy + y^2$.

Solution   We have the squares x^2 and $(-y)^2$, and the middle term $2x(-y)$. Therefore,

$$x^2 - 2xy + y^2 = x^2 + 2x(-y) + (-y)^2$$
$$= (x - y)^2.$$

To check, we multiply:

$$(x - y)^2 = (x - y)(x - y)$$
$$= x^2 - xy - xy + y^2$$
$$= x^2 - 2xy + y^2.$$

We may find perfect squares involving squares of squares, squares of cubes, and so on.

EXAMPLE 17.15   Factor $9x^6 - 24x^3y^3 + 16y^6$.

Solution   We observe that $9x^6 = (3x^3)^2$ and $16y^6 = (-4y^3)^2$. Also, the middle term is $-24x^3y^3 = 2(3x^3)(-4y^3)$. Therefore, we have a perfect square:

$$9x^6 - 24x^3y^3 + 16y^4 = (3x^3)^2 + 2(3x^3)(-4y^3) + (-4y^3)^2$$
$$= (3x^3 - 4y^3)^2.$$

To check, we multiply:

$$(3x^3 - 4y^3)^2 = (3x^3 - 4y^3)(3x^3 - 4y^3)$$
$$= (3x^3)(3x^3) - (3x^3)(4y^3) - (4y^3)(3x^3) + (4y^3)(4y^3)$$
$$= 9x^6 - 12x^3y^3 - 12x^3y^3 + 16y^6$$
$$= 9x^6 - 24x^3y^3 + 16y^6.$$

As before, we factor out any common factors before looking for a perfect square.

EXAMPLE 17.16   Factor $4x^4 - 16x^3 + 16x^2$.

Solution   There is a common factor, $4x^2$. We factor out the common factor and then factor the resulting perfect square:

$$4x^4 - 16x^3 + 16x^2 = 4x^2(x^2 - 4x + 4)$$
$$= 4x^2(x - 2)^2.$$

To check, you should first multiply the perfect square, and then multiply by the monomial.

Exercise 17.3

Factor:

1. $x^2 + 2x + 1$

2. $x^2 + 8x + 16$

3. $x^2 - 6x + 9$

4. $x^2 - 16x + 64$

5. $4x^2 - 20x + 25$

6. $25x^2 - 40x + 16$

7. $4x^2 + 4x + 1$

8. $81x^2 - 18x + 1$

9. $x^2 + 16xy + 64y^2$

10. $16x^2 - 24xy + 9y^2$

11. $x^4 - 4x^2 + 4$

12. $49x^4 + 14x^2 + 1$

13. $x^4 + 2x^2y^2 + y^4$

14. $9x^6 - 12x^3y^3 + 4y^6$

15. $4x^2 - 16x + 16$

16. $4x^2y^2 - 4xy^2 + y^2$

17. $4x^4 + 24x^2y^2 + 36y^4$

18. $36x^4 - 216x^3 + 324x^2$

19. $x^8 + 12x^6 + 36x^4$

20. $16x^8 - 64x^4y^2 + 64y^4$

17.4   The General Trinomial

Many trinomials other than perfect squares can be factored over the integers. Consider, for example, the trinomial

$$x^2 + 3x + 2.$$

We recall from Section 15.3 that if this trinomial is the product of two binomials, then

x^2 is the product of the first terms,
2 is the product of the last terms, and
$3x$, the middle term, is the result of combining the products of the outside and inside terms.

Since all the terms of the trinomial have positive coefficients, all the terms of any possible binomial factors have positive coefficients. The only positive factors of 2 are 1 and 2. Therefore, the only possible binomial factorization is $(x + 1)(x + 2)$. This factorization gives a middle term of $3x = 2x + x$; therefore,

$$x^2 + 3x + 2 = (x + 1)(x + 2).$$

The factorization can be checked by multiplying the two binomial factors.

Not all trinomials can be factored into two binomial factors. Consider, for example, the trinomial

$$x^2 + x + 1.$$

Again, all of the coefficients are positive. The only positive factors of 1 are 1 and 1. But we know that

$$(x + 1)(x + 1) = x^2 + 2x + 1.$$

Since there are no other possible factors, we conclude that the trinomial $x^2 + x + 1$ cannot be factored over the integers.

For the rest of the unit we will consider only trinomials that can be factored into two binomial factors over the integers.

EXAMPLE 17.17   Factor $x^2 + 5x + 6$.

Solution   The first terms must each be x. Therefore, we write

$$x^2 + 5x + 6 = (x \qquad)(x \qquad).$$

Also, all of the coefficients are positive. The positive factors of 6 are 1 and 6 or 2 and 3. We try the posssible products;

$$(x + 1)(x + 6) = x^2 + 7x + 6$$

and

$$(x + 2)(x + 3) = x^2 + 5x + 6.$$

The middle term of the original trinomial is $5x = 3x + 2x$; therefore,

$$x^2 + 5x + 6 = (x + 2)(x + 3).$$

If the coefficient of the middle term of a trinomial is negative, but the constant term is positive, the constant term is a product of negative integers.

EXAMPLE 17.18   Factor $x^2 - 9x + 8$.

Solution   Again, we write

$$x^2 - 9x + 8 = (x \qquad)(x \qquad).$$

Here the coefficient of the middle term is negative and the constant term is positive. The negative factors of 8 are -1 and -8 or -2 and -4. We try the possible products

$$(x - 1)(x - 8) = x^2 - 9x + 8$$

and

$$(x - 2)(x - 4) = x^2 - 6x + 8.$$

The middle term of the original trinomial is $-9x = -8x - x$; therefore,

$$x^2 - 9x + 8 = (x - 1)(x - 8).$$

If the constant term of a trinomial is negative, it is a product of one positive factor and one negative factor.

EXAMPLE 17.19   Factor $x^2 + x - 6$.

Solution   We write

$$x^2 + x - 6 = (x \qquad)(x \qquad).$$

Since the constant term of the trinomial is negative, its factors could be 1 and -6, -1 and 6, 2 and -3, or -2 and 3. We rule out the cases where the larger number has the negative sign, because in these cases the middle term will be negative. Then, since $x = 3x - 2x$,

$$x^2 + x - 6 = (x - 2)(x + 3).$$

You should check the factorization by multiplying.

When the coefficient of the first term of a trinomial is not 1, we must consider its effect on the middle term.

EXAMPLE 17.20   Factor $2x^2 - 3x - 2$.

Solution   Since $2x^2 = 2x \cdot x$, we write

$$2x^2 - 3x - 2 = (2x \qquad)(x \qquad).$$

The constant term of the trinomial is negative, so the constant terms of the binomial factors could be -1 and 2 or 1 and -2. Suppose we try -1 and 2:

$$(2x - 1)(x + 2) = 2x^2 + 3x - 2.$$

The factorization is not correct. Since the first terms of the binomial factors are different, we must also try -1 and 2 in reverse order:

$$(2x + 2)(x - 1) = 2x^2 - 2.$$

We now try 1 and -2:

$$(2x + 1)(x - 2) = 2x^2 - 3x - 2.$$

This factorization gives the original trinomial. Therefore,

$$2x^2 - 3x - 2 = (2x + 1)(x - 2).$$

Had this combination not worked, we would have tried 1 and -2 in reverse order. When a trinomial has several possible combinations, you will have to do some searching by trial and error. It helps to be able to find the middle term mentally, without writing out the entire multiplication of the possible factors you are considering.

EXAMPLE 17.21   Factor $4x^2 - 19x + 12$.

Solution   The first terms may be $2x$ and $2x$ or $4x$ and x. The constant terms must both be negative, and could be -1 and -12, -2 and -6, or -3 and -4. We might try

$$4x^2 - 19x + 12 = (2x \qquad)(2x \qquad).$$

Trying each of the possible pairs of second terms, we find

$(2x - 1)(2x - 12)$ has middle term $-26x$,
$(2x - 2)(2x - 6)$ has middle term $-16x$, and
$(2x - 3)(2x - 4)$ has middle term $-14x$.

Since none of these possibilities gives the correct middle term, we try

$$4x^2 - 19x + 12 = (4x \qquad)(x \qquad).$$

Now, since the first terms are different, we must try each possible pair of constants both ways. Thus,

$(4x - 1)(x - 12)$ has middle term $-49x$, and
$(4x - 12)(x - 1)$ has middle term $-16x$,

so the last terms are not -1 and 12.

$(4x - 2)(x - 6)$ has middle term $-26x$, and
$(4x - 6)(x - 2)$ has middle term $-14x$,

so the last terms are not -2 and -6. However,

$(4x - 3)(x - 4)$ has middle term $-19x$, and therefore

$$4x^2 - 19x + 12 = (4x - 3)(x - 4).$$

To be sure there are no mistakes, we recheck by multiplying:

$$(4x - 3)(x - 4) = 4x^2 - 16x - 3x + 12$$
$$= 4x^2 - 19x + 12.$$

EXAMPLE 17.22   Factor $6x^2 + x - 15$.

Solution   The first terms may be $2x$ and $3x$ or $6x$ and x. The constant terms must have opposite signs. They could be 1 and -15, -1 and 15, 3 and -5, or -3 and 5. We might try

$$6x^2 + x - 15 = (2x \qquad)(3x \qquad).$$

Trying each of the possible pairs of constants, we find that the first three pairs all give the wrong middle term. However, $(2x - 3)(3x + 5)$ has middle term x, and so

$$6x^2 + x - 15 = (2x - 3)(3x + 5).$$

The trinomial is factored, and we need not try $6x$ and x as first terms. Remember to recheck by multiplying.

A trinomial may be a perfect square. If we do not recognize the perfect square as the form shown in Section 17.3, we will find the factorization anyway, by trial and error.

EXAMPLE 17.23   Factor $9x^2 - 12x + 4$.

Solution   The first terms may be $3x$ and $3x$ or $9x$ and x. The constant terms must both be negative. They may be -1 and -4 or -2 and -2. We might try

$$9x^2 - 12x + 4 = (3x \qquad)(3x \qquad).$$

Trying the possible pairs of constant terms, we find that $(3x - 2)(3x - 2)$ has middle term $-12x$, and so

$$9x^2 - 12x + 4 = (3x - 2)(3x - 2)$$
$$= (3x - 2)^2.$$

You should recheck by multiplying.

A trinomial may look like a perfect square and yet not be.

EXAMPLE 17.24   Factor $9x^2 - 20xy + 4y^2$.

Solution   Again, the first terms may be $3x$ and $3x$ or $9x$ and x. The second terms may be $-y$ and $-4y$ or $-2y$ and $-2y$. Again, trying

$$9x^2 - 20xy + 4y^2 = (3x \qquad)(3x \qquad),$$

we find

$(3x - y)(3x - 4y)$ has middle term $-15xy$, and
$(3x - 2y)(3x - 2y)$ has middle term $-12xy$.

These possibilities do not give the correct middle term. Therefore, we try

$$9x^2 - 20xy + 4y^2 = (9x \qquad)(x \qquad).$$

Since the first terms are different, we must try each possible pair of second terms two ways:

$(9x - y)(x - 4y)$ has middle term $-14xy$, and
$(9x - 4y)(x - y)$ has middle term $-13xy$,

so the last terms are not $-y$ and $-4y$. Next, however, we find

$(9x - 2y)(x - 2y)$ has middle term $-20xy$.

Thus,

$$9x^2 - 20xy + 4y^2 = (9x - 2y)(x - 2y).$$

You should recheck by multiplying.

We always factor out any common factors before factoring a trinomial.

EXAMPLE 17.25   Factor $4x^2 - 16x + 12$.

Solution   We factor out the common factor 4:

$$4x^2 - 16x + 12 = 4(x^2 - 4x + 3).$$

The remaining trinomial is easily factored. The first terms are both x, and the second terms must be -1 and -3:

$$x^2 - 4x + 3 = (x - 1)(x - 3).$$

Thus,

$$4x^2 - 16x + 12 = 4(x^2 - 4x + 3)$$
$$= 4(x - 1)(x - 3).$$

To check, you should multiply the two binomial factors, and then multiply by the monomial.

Exercise 17.4

Factor:

1. $x^2 + 5x + 4$ 2. $x^2 + 9x + 18$ 3. $x^2 - 7x + 12$ 4. $x^2 - 10x + 16$

5. $x^2 - 2x - 15$ 6. $x^2 + 2x - 48$ 7. $2x^2 + x - 6$ 8. $3x^2 - 16x + 5$

9. $4x^2 + 13x - 35$ 10. $6x^2 + 19x - 20$ 11. $6x^2 + 25x + 24$ 12. $8x^2 - 2x - 45$

13. $9x^2 - 30x + 16$ 14. $8x^2 - 75x - 50$ 15. $x^2 + 12x + 36$ 16. $4x^2 - 12x + 9$

17. $4x^2 - 15x + 9$ 18. $25x^2 + 25x + 4$ 19. $x^2 - 12xy + 36y^2$ 20. $x^2 - 15xy + 36y^2$

21. $3x^2 + 12x + 9$ 22. $10x^2 + 10x - 60$ 23. $12x^2 - 44x + 24$ 24. $10x^2 - 25x - 60$

25. $x^4 + 3x^3 + 2x^2$ 26. $8x^4 + 28x^3 - 16x^2$ 27. $x^2y^2 - xy^2 - 6y^2$ 28. $x^3y + 2x^2y^2 + xy^3$

29. $x^4 - 5x^2 + 6$ 30. $2x^4 - 5x^2y^2 + 2y^4$

Self-test

Factor:

1. $25x^2 - 60x + 36$

2. $5x^2 - 25x + 30$

3. $3x^2 - 9x + 27$

4. $9x^2 - 25y^2$

5. $2x^2 + xy - 6y^2$

1. _____

2. _____

3. _____

4. _____

5. _____

UNIT 18 Solving Equations by Factoring

INTRODUCTION

In the preceding unit you learned how to rewrite certain nonlinear expressions as products of linear expressions by factoring. Using these methods, you can solve certain equations that are not linear by reducing them to linear equations. In this unit you will learn how to solve two types of nonlinear equations. You will also learn two applications of solving such equations.

OBJECTIVES

When you have finished this unit you should be able to:

1. Solve equations that have a variable common factor.
2. Solve quadratic equations by factoring.
3. Solve problems involving rectangular areas using quadratic equations that can be solved by factoring.
4. Solve problems involving the Pythagorean theorem.

18.1 Variable Common Factors

Although an equation may be simplified by dividing each term of the equation by a *constant* factor, we may not divide each term of an equation by a *variable* factor. For example, consider the equation

$$x^2 = x.$$

It is tempting to divide by x and find $x = 1$. This result is in fact a solution of the equation, since

$$1^2 = 1.$$

It is possible, however, that the variable x is zero. If this is the case, we have divided by zero, which is not allowed. Indeed, substituting $x = 0$ into the original equation gives

$$0^2 = 0.$$

Thus, $x = 0$ is another solution to the equation. In dividing by x we divided by zero, and in doing so we lost the solution $x = 0$.

Since we cannot divide an equation by a variable factor, we must use another method for solving equations like the one above. We use a theorem that we will call the product rule.

Product Rule: If $ab = 0$, then $a = 0$ or $b = 0$ or both.

The proof of the product rule is given at the end of this unit.

247

To use the product rule to solve an equation such as $x^2 = x$, we collect the nonzero terms on one side, leaving zero on the other side, and then factor out the common factor.

EXAMPLE 18.1   Solve $x^2 = x$.

Solution   Collecting the terms on the left-hand side, we have

$$x^2 = x$$
$$x^2 - x = 0.$$

Now we factor out the common factor x:

$$x(x - 1) = 0.$$

Then, using the product rule,

$$x = 0 \text{ and } x - 1 = 0$$
$$x = 0 \text{ and } x = 1.$$

Since $x = 0$ and $x = 1$ each make the original equation true, both 0 and 1 are solutions of the equation.

EXAMPLE 18.2   Solve $2x^2 + 4x = 0$.

Solution   We may remove the constant factor 2 by dividing every term of the equation by 2:

$$\frac{2x^2}{2} + \frac{4x}{2} = \frac{0}{2}$$
$$x^2 + 2x = 0.$$

However, we must not divide by x, so we continue by factoring and using the product rule:

$$x(x + 2) = 0$$
$$x = 0 \text{ and } x + 2 = 0$$
$$x = 0 \text{ and } x = -2.$$

The solutions are 0 and -2. As always, you should check these solutions in the original equation.

When we use the product rule, we must collect all like terms on one side of the equation before factoring.

EXAMPLE 18.3   Solve $3x^2 - 4x = x^2 + x$.

Solution   We collect the terms on the left-hand side of the equation:

$$3x^2 - 4x = x^2 + x$$
$$2x^2 - 5x = 0$$
$$x(2x - 5) = 0$$
$$x = 0 \text{ and } 2x - 5 = 0$$
$$x = 0 \text{ and } x = \frac{5}{2}.$$

The solutions are 0 and $\frac{5}{2}$.

If necessary, we simplify all expressions in the equation, collect all like terms on one side, and then use the product rule.

EXAMPLE 18.4 Solve $x(x - 4) = x(2x - 3)$.

Solution   First, we simplify the expressions on each side of the equation:

$$x(x - 4) = x(2x - 3)$$
$$x^2 - 4x = 2x^2 - 3x.$$

Remember that we cannot divide by x at any time. Now, we combine like terms and use the product rule. We collect the terms on the right-hand side of the equation:

$$x^2 - 4x = 2x^2 - 3x$$
$$0 = x^2 + x$$
$$0 = x(x + 1)$$
$$x = 0 \text{ and } x + 1 = 0$$
$$x = 0 \text{ and } x = -1.$$

The solutions are 0 and -1.

Exercise 18.1

Solve and check:

1. $x^2 = 5x$ 2. $2x^2 = x$

3. $2x^2 = 6x$ 4. $3x^2 + 9x = 0$

5. $4x^2 + 6x = 0$ 6. $5x = 15x^2$

7. $4x^2 + 3x = x^2 - 2x$ 8. $x^2 - 6x = 3x^2 + 2x$

9. $x(3x - 5) = x(2x + 1)$ 10. $2x(2x + 3) = 6x(x + 4)$

18.2 Solving Quadratic Equations by Factoring

The algebraic expressions making up a **quadratic equation** contain variables to the second power, or **second-degree terms.** They may also contain variables to the first power, or **first-degree terms,** and constants, or **constant terms.** The equations in Section 18.1 are quadratic equations. Some other examples of quadratic equations are:

$$x^2 - 3x + 2 = 0, \quad 2x^2 - 8 = 0, \quad x^2 = 0.$$

Observe that, although there must be a second-degree term in a quadratic equation, there need not be a first-degree term or a constant term.

To solve the equations in Section 18.1, we collected all the nonzero terms on one side, factored, and used the product rule. Because these equations contained no constant term, factoring was a matter of factoring out a variable common factor. The product rule can also be used for quadratic equations that do have a constant term, if the algebraic expression consisting of all the nonzero terms can be factored by the methods in Unit 17.

EXAMPLE 18.5   Solve $x^2 = 9$.

Solution   First, we collect the nonzero terms:

$$x^2 = 9$$
$$x^2 - 9 = 0.$$

Factoring and using the product rule, we have

$$(x + 3)(x - 3) = 0$$
$$x + 3 = 0 \text{ and } x - 3 = 0$$
$$x = -3 \text{ and } x = 3.$$

Both -3 and 3 are solutions. To check, for $x = -3$,

$$(-3)^2 = 9,$$

and for $x = 3$,

$$3^2 = 9.$$

EXAMPLE 18.6   Solve $x^2 - 3x + 2 = 0$.

Solution   Since all the nonzero terms are on the left-hand side, we try to factor the trinomial into two binomial factors. This trinomial factors easily, and we have

$$x^2 - 3x + 2 = 0$$
$$(x - 1)(x - 2) = 0.$$

Now, using the product rule,

$$x - 1 = 0 \text{ and } x - 2 = 0$$
$$x = 1 \text{ and } x = 2.$$

Both 1 and 2 are solutions, as you can verify by checking in the original equation.

If all the nonzero terms are not on one side of a quadratic equation, we must collect them before we factor. This step is necessary because we must have a zero on one side of the equation in order to use the product rule. If neither side of an equation is zero, the product rule does not hold. For example, if $ab = 1$, then it is not necessarily true that $a = 1$ or $b = 1$ or both. We could have $a = -1$ and $b = -1$, or $a = 2$ and $b = \frac{1}{2}$, and so on. Thus, to use the product rule, we must always write an equation with all the nonzero terms on one side and zero on the other.

EXAMPLE 18.7   Solve $2x^2 - x = 1$.

Solution   First, we collect the nonzero terms:

$$2x^2 - x = 1$$
$$2x^2 - x - 1 = 0.$$

Now we may factor the trinomial and use the product rule to complete the solution:

$$(x - 1)(2x + 1) = 0$$
$$x - 1 = 0 \text{ and } 2x + 1 = 0$$
$$x = 1 \text{ and } x = -\frac{1}{2}.$$

The solutions are 1 and $-\frac{1}{2}$. You should check these solutions in the original equation.

The x^2 term in the trinomials we have factored has been positive. We may collect terms on the right-hand side of an equation to produce a positive x^2 term.

EXAMPLE 18.8   Solve $x + 1 = 6x^2$.

Solution   We collect on the right so that the x^2 term is positive:

$$x + 1 = 6x^2$$
$$0 = 6x^2 - x - 1$$
$$0 = (2x - 1)(3x + 1)$$
$$2x - 1 = 0 \text{ and } 3x + 1 = 0$$
$$2x = 1 \text{ and } 3x = -1$$
$$x = \frac{1}{2} \text{ and } x = -\frac{1}{3}.$$

The solutions are $\frac{1}{2}$ and $-\frac{1}{3}$.

Although we must never divide out a term involving a variable, we may divide out any constant common factors.

EXAMPLE 18.9   Solve $2x^2 - 8 = 0$.

Solution   Since 2 is a constant, we may divide by 2:

$$2x^2 - 8 = 0$$
$$\frac{2x^2}{2} - \frac{8}{2} = \frac{0}{2}$$
$$x^2 - 4 = 0.$$

Factoring and using the product rule,

$$(x + 2)(x - 2) = 0$$
$$x + 2 = 0 \text{ and } x - 2 = 0$$
$$x = -2 \text{ and } x = 2.$$

The solutions are -2 and 2.

In the special case where the trinomial is a perfect square, the factored expression has two identical factors.

EXAMPLE 18.10   Solve $4x = x^2 + 4$.

Solution   Collecting on the right, we have

$$4x = x^2 + 4$$
$$0 = x^2 - 4x + 4$$
$$0 = (x - 2)(x - 2)$$
$$x - 2 = 0$$
$$x = 2.$$

Since this equation has identical factors, we need only write the repeated factor once, and we only write the solution derived from the factor once. However, it is sometimes convenient to think of an equation with two identical factors as having two identical solutions. For example, the equation above has two identical solutions, each of which is 2. This type of solution is often called a "double root."

When solving a quadratic equation, as with other equations, we first simplify all expressions. We then collect terms on one side and use the product rule.

EXAMPLE 18.11   Solve $3x(x + 1) = 10 - (1 - x^2)$.

Solution   First, we simplify the expressions on each side:

$$3x(x + 1) = 10 - (1 - x^2)$$
$$3x^2 + 3x = 10 - 1 + x^2$$
$$3x^2 + 3x = 9 + x^2.$$

Then, we combine like terms on one side and use the product rule:

$$2x^2 + 3x - 9 = 0$$
$$(2x - 3)(x + 3) = 0$$
$$2x - 3 = 0 \text{ and } x + 3 = 0$$
$$2x = 3 \text{ and } x = -3$$
$$x = \frac{3}{2} \text{ and } x = -3.$$

The solutions are $\frac{3}{2}$ and -3. When you check these solutions, be sure to use the original equation.

Exercise 18.2

Solve and check:

1. $x^2 - 1 = 0$ 2. $x^2 = 25$

3. $16x^2 = 1$ 4. $9x^2 - 4 = 0$

5. $x^2 - 5x + 6 = 0$ 6. $x^2 + 4x + 3 = 0$

7. $x^2 - 2x = 15$ 8. $x^2 + 4x = 60$

9. $3x^2 - x = 2$ 10. $4x^2 + x = 3$

11. $4x + 3 = 4x^2$ 12. $26x - 5 = 5x^2$

13. $5x^2 - 80 = 0$ 14. $3x^2 = 6 - 3x$

15. $2x^2 = x$ 16. $3x^2 + 9x = 0$

17. $x^2 + 6x + 9 = 0$ 18. $2x^2 + 2 = 4x$

19. $14x - 4x^2 = 8 - x^2$ 20. $6x^2 + 7x = 4x + 3$

21. $2(x - 4) = x(x + 8)$ 22. $1 - (3 - x^2) = 7x(x - 1)$

23. $(x - 2)(x - 7) = 50$ 24. $(x + 2)^2 = 1$

18.3   Rectangular Areas

Problems involving areas of geometric figures provide an application of solving quadratic equations. Recall from Section 8.2 the description of a rectangle. For a rectangle

w [ ] l

if w is the width and l is the length, the area is

$$A = wl,$$

or the width times the length.

EXAMPLE 18.12   A rectangle has a length of 8 inches more than its width. Its area is 33 square inches. Find the width and the length.

Solution   The length is 8 inches more than the width; that is, if the width is w, the length is $w + 8$. A diagram of the rectangle is

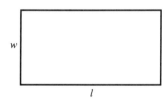

w $A = 33$

$w + 8$

The area is given by the width times the length, so we write

$$wl = A$$

or

$$w(w + 8) = 33.$$

We remove the parentheses on the left-hand side and collect terms:

$$w^2 + 8w = 33$$
$$w^2 + 8w - 33 = 0.$$

Then, we factor and solve for w:

$$(w + 11)(w - 3) = 0$$
$$w + 11 = 0 \text{ and } w - 3 = 0$$
$$w = -11 \text{ and } w = 3.$$

The solution $w = 3$ gives a width of 3 inches. Since the length is 8 more, the length is $3 + 8 = 11$ inches. To check, we observe that the area is $(3)(11) = 33$ square inches.

Since a side of a rectangle cannot be negative, the solution -11 is meaningless. We call such a solution an **extraneous solution.** An extraneous solution is one that is derived by correct methods, but does not fit the special conditions of a problem.

EXAMPLE 18.13   The total area of this diagram is 80 square inches:

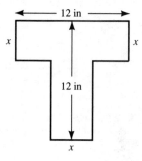

Find the width x.

Solution   We divide the diagram into two rectangles as shown by the dotted line:

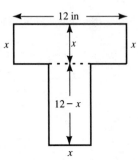

The width of the top rectangle is x and its length is 12. Therefore, the area of the top rectangle is

$$12x.$$

We have cut off x inches from the height of the figure to create the bottom rectangle. Thus, the length of the bottom rectangle is $12 - x$. Its width is x. Therefore, its area is

$$x(12 - x).$$

The total area of the figure is the sum of the areas of the two separate rectangles:

$$12x + x(12 - x) = 80.$$

We simplify the left-hand side of this equation, collect terms, and factor to solve the equation:

$$12x + 12x - x^2 = 80$$
$$24x - x^2 = 80$$
$$0 = x^2 - 24x + 80$$
$$0 = (x - 4)(x - 20)$$
$$x - 4 = 0 \text{ and } x - 20 = 0$$
$$x = 4 \text{ and } x = 20.$$

The width of the figure is 4 inches. You can check this solution by finding the area of each rectangle and the total area. In this case, $x = 20$ is an extraneous solution because the length of the rectangle would be $12 - 20$, which is impossible.

EXAMPLE 18.14   A rectangular garden is 11 feet by 15 feet. A walkway of uniform width is laid around the garden, and the area of the resulting rectangle is 396 square feet. How wide is the walkway?

Solution   We carefully draw a diagram, showing the garden and the walkway. We have labeled the width of the walkway x:

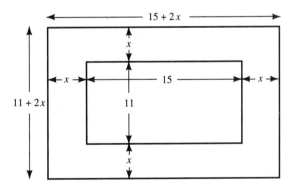

Adding the width of the walkway on each side, the width of the resulting rectangle is $11 + 2x$, and the length is $15 + 2x$. Therefore, the area of the resulting rectangle is

$$(11 + 2x)(15 + 2x) = 396.$$

We find the product on the left-hand side and collect terms:

$$165 + 52x + 4x^2 = 396$$
$$4x^2 + 52x - 231 = 0.$$

It will take some experimentation to factor the left-hand side. The factors of 231 are 1 and 231, 3 and 77, 7 and 33, and 11 and 21. The correct factorization gives

$$(2x + 33)(2x - 7) = 0$$
$$2x + 33 = 0 \text{ and } 2x - 7 = 0$$
$$2x = -33 \text{ and } 2x = 7$$
$$x = -\frac{33}{2} \text{ and } x = \frac{7}{2}.$$

The width of the walkway is $\frac{7}{2}$ or $3\frac{1}{2}$ feet. To check, we observe that the width of the outer rectangle is then $3\frac{1}{2} + 11 + 3\frac{1}{2} = 18$ feet, and its length is $3\frac{1}{2} + 15 + 3\frac{1}{2} = 22$ feet. Its area is then $(18)(22) = 396$ square feet. The negative solution is extraneous.

Exercise 18.3

1. The length of a rectangle is 3 centimeters more than its width. The area is 70 square centimeters. Find the width and the length.

2. A rectangle has an area of 220 square inches. The length is 9 inches more than the width. Find the width and the length.

3. The width of a rectangle is 15 feet less than its length. The area is 450 square feet. Find the length and the width.

4. The width of a rectangle is $2\frac{1}{2}$ meters less than its length. The area is $37\frac{1}{2}$ square meters. Find the length and the width.

5. The total area of this diagram is 176 square inches. Find the width x.

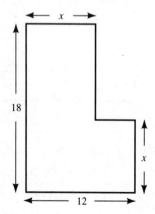

6. Find the width x of this diagram, if the area is 72 square centimeters.

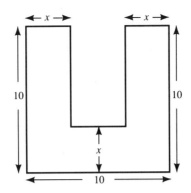

7. A rectangular swimming pool is 10 meters by 25 meters. There is a concrete edge of uniform width around the pool. If the area of the resulting rectangle is 700 square meters, how wide is the concrete edge?

8. A rectangular garden has a fence at the back. A border of uniform width is put along the two shorter sides and the front. If the garden is 5 feet by 12 feet without the border, and the rectangle produced by adding the border has an area of 112 square feet, how wide is the border?

18.4   The Pythagorean Theorem

We have used squares of numbers from Unit 14 of this book onward. In other courses, you may also have used the **square roots** of numbers. For example, the positive square root of 4 is written $\sqrt{4}$. We say that $\sqrt{4}$ is the positive number a such that $4 = a^2$. That is,

$$\sqrt{4} = a \text{ if } 4 = a^2.$$

But we know that $4 = 2^2$. Therefore,

$$\sqrt{4} = 2 \text{ because } 4 = 2^2.$$

Similarly,

$$\sqrt{9} = 3 \text{ because } 9 = 3^2,$$
$$\sqrt{16} = 4 \text{ because } 16 = 4^2,$$
$$\sqrt{25} = 5 \text{ because } 25 = 5^2,$$

and so on. Also,

$$\sqrt{1} = 1 \text{ because } 1 = 1^2$$

and

$$\sqrt{100} = 10 \text{ because } 100 = 10^2.$$

We may have to search to find square roots of larger numbers. For example, to find $\sqrt{256}$, we start with numbers greater than 10. We know that $\sqrt{100} = 10$ and that 256 is more than 100. Also,

$\sqrt{256}$ is an even number, because an odd number times itself always gives an odd number. Trying even numbers greater than 10, we find

$$\sqrt{256} = 16 \text{ because } 256 = 16^2.$$

We will use square roots in another application of factoring.

A **right triangle** is a triangle in which one of the angles is a **right angle.** The side of the triangle opposite the right angle is called the **hypotenuse:**

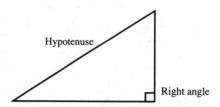

A famous property of right triangles is named for the ancient Greek mathematician Pythagoras (c. 540 B.C., although it was known to ancient cultures more than 1000 years earlier). We state the **Pythagorean theorem** without proof.

Pythagorean Theorem: If a, b, and c are sides of a right triangle, with c the hypotenuse, then

$$c^2 = a^2 + b^2.$$

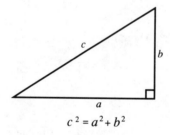

$$c^2 = a^2 + b^2$$

EXAMPLE 18.15   Two sides of a right triangle are 6 inches and 8 inches. Find the hypotenuse.

Solution   A diagram of the triangle is

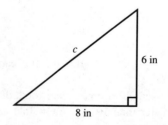

Using the Pythagorean theorem,

$$c^2 = a^2 + b^2$$
$$c^2 = 6^2 + 8^2$$
$$c^2 = 36 + 64$$
$$c^2 = 100.$$

It is not necessary to solve this equation by factoring. If

$$c^2 = 100$$

then

$$c = \sqrt{100}$$
$$c = 10.$$

The hypotenuse is 10 inches.

If the hypotenuse and a side of a right triangle are given, and the remaining side is to be found, we can avoid large numbers by using the difference of squares.

EXAMPLE 18.16   The diagonal of a rectangle is 34 centimeters. Its width is 16 centimeters. Find its length.

Solution   The diagonal of a rectangle is a line joining two opposite corners, forming two right triangles. A diagram is

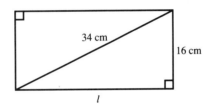

Using the Pythagorean theorem,

$$c^2 = a^2 + b^2$$
$$34^2 = l^2 + 16^2$$
$$34^2 - 16^2 = l^2$$

or

$$l^2 = 34^2 - 16^2.$$

The right-hand side is a difference of squares, which we can factor:

$$l^2 = 34^2 - 16^2$$
$$l^2 = (34 + 16)(34 - 16)$$
$$l^2 = (50)(18)$$
$$l^2 = 900.$$

Therefore,

$$l = \sqrt{900}$$
$$l = 30.$$

The length is 30 centimeters. To check, observe that

$$34^2 \overset{?}{=} 30^2 + 16^2$$
$$1156 \overset{?}{=} 900 + 256$$
$$1156 = 1156.$$

EXAMPLE 18.17   A pole is supported by a guy wire 13 feet long. One end of the wire is anchored to the ground 5 feet from the base of the pole. How far up the pole does the wire reach?

Solution   Assuming that the pole is vertical and the ground is horizontal, the pole and the ground form the sides of a right triangle. The wire is the hypotenuse. A diagram is

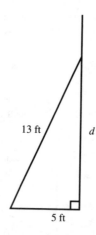

Using the Pythagorean theorem,

$$c^2 = a^2 + b^2$$
$$13^2 = 5^2 + d^2$$
$$13^2 - 5^2 = d^2.$$

Thus,

$$d^2 = 13^2 - 5^2$$
$$d^2 = (13 + 5)(13 - 5)$$
$$d^2 = (18)(8)$$
$$d^2 = 144$$
$$d = \sqrt{144}$$
$$d = 12.$$

The distance is 12 feet. To check,

$$13^2 \overset{?}{=} 5^2 + 12^2$$
$$169 \overset{?}{=} 25 + 144$$
$$169 = 169.$$

Exercise 18.4

1. Two sides of a right triangle are 3 feet and 4 feet. Find the hypotenuse.

2. The width and length of a rectangle are 8 meters and 15 meters. Find the length of the diagonal.

3. The hypotenuse of a right triangle is 39 centimeters. One side is 36 centimeters. Find the other side.

4. The diagonal of a rectangle is 40 inches. The length is 32 inches. Find the width.

5. A pole is supported by a guy wire 50 inches long. The wire reaches 48 inches up the pole. How far from the base of the pole is the end of the wire at the ground?

6. A ladder 41 feet long reaches 40 feet up the side of a house. How far from the base of the house is the foot of the ladder?

7. A diagonal support 29 centimeters long is placed under a shelf. One end is at the wall, 21 centimeters below the shelf. If the other end is at the edge of the shelf, how far out does the shelf extend?

8. The diagonal distance from a point at the top of a cliff to a boat is 109 meters. The point at the top of the cliff is 91 meters above the water. Assuming the cliff is vertical, how far is the boat from the base of the cliff?

Proof of the Product Rule:
Suppose $ab = 0$. Consider the factor a. Either $a = 0$ or $a \neq 0$. Thus we have two cases: (1) $a = 0$ and (2) $a \neq 0$.
Case 1. If $a = 0$, then $a = 0$ or $b = 0$ or both, and the theorem is proved for this case.
Case 2. If $a \neq 0$, then we can divide by a since we will not be dividing by zero:

$$\frac{ab}{a} = \frac{0}{a}$$
$$b = 0.$$

Thus $a = 0$ or $b = 0$ or both, and the theorem is proved for this case.

Self-test

Solve by factoring and check:

1. $x(x + 6) = 9(x + 2)$

1. _____

2. $2x^2 - 12x = 0$

2. _____

3. $2x^2 - 7x = 4$

3. _____

4. A ladder is placed so that its foot is 10 feet from the base of a house, and it reaches 24 feet up the side of the house. How long is the ladder?

4. _____

5. Find the width x in this diagram if the area is 180 square centimeters:

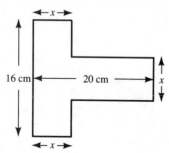

5. _____

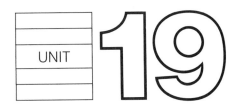

Rational Expressions

INTRODUCTION

In the two preceding units you learned how to factor polynomials and how to apply factoring to solving quadratic equations. In this unit you will learn how to use factoring to simplify rational expressions. You will reduce rational expressions, and multiply and divide them. Then, you will learn how to use a common denominator to add and subtract rational expressions.

OBJECTIVES

When you have finished this unit you should be able to:

1. Reduce a rational expression to lowest terms.
2. Multiply and divide rational expressions, and reduce to lowest terms.
3. Use the least common denominator to add and subtract rational expressions.

19.1

Reducing Rational Expressions

A **rational number** is a quotient, or ratio, of two integers, in which the denominator is not zero. Thus, if a and b are integers, with $b \neq 0$, then $\dfrac{a}{b}$ is a rational number. Some examples of rational numbers are

$$\frac{2}{3}, \quad \frac{-5}{6}, \quad \frac{0}{1}.$$

A **rational expression** is a quotient, or ratio, of two polynomials, in which the denominator is not zero. Thus, if A and B are polynomials, with $B \neq 0$, then $\dfrac{A}{B}$ is a rational expression. Some examples of rational expressions are

$$\frac{3x + 6}{5}, \quad \frac{x - 2}{x + 3}, \quad \frac{x^2 - 2x + 1}{x^2 - 1}.$$

Recall that we can reduce rational numbers by dividing out a common factor from the numerator and the denominator. For example,

$$\frac{3}{6} = \frac{3(1)}{3(2)} = \frac{3}{3}\left(\frac{1}{2}\right) = 1\left(\frac{1}{2}\right) = \frac{1}{2}.$$

Similarly, we can reduce a rational expression by dividing out a common factor from the numerator and the denominator.

EXAMPLE 19.1 Reduce $\dfrac{3x}{6}$.

Solution   We find a common factor 3 in the numerator and the denominator. Thus,

$$\frac{3x}{6} = \frac{3x}{3(2)}$$

$$= \frac{3}{3}\left(\frac{x}{2}\right).$$

Then, since $\frac{3}{3} = 1$,

$$\frac{3}{3}\left(\frac{x}{2}\right) = 1\left(\frac{x}{2}\right)$$

$$= \frac{x}{2}.$$

A common factor in the numerator and the denominator of a rational expression may contain variables as well as constants.

| EXAMPLE 19.2   Reduce $\dfrac{2x}{4x^2}$.

Solution   There is a common factor 2 in the numerator and the denominator. Also, there is a common factor x. Thus,

$$\frac{2x}{4x^2} = \frac{2x}{(2)(2)(x)(x)}$$

$$= \frac{2x}{2x}\left(\frac{1}{2x}\right)$$

$$= 1\left(\frac{1}{2x}\right)$$

$$= \frac{1}{2x}.$$

Observe that all of the factors in the numerator seem to have been divided out. However, a 1 is left in the numerator because $2x = 2x(1)$.

We may use the techniques of factoring to reduce rational expressions.

| EXAMPLE 19.3   Reduce $\dfrac{3 + 12x}{3}$.

Solution   We use the distributive property to factor out the common factor 3 in the numerator:

$$\frac{3 + 12x}{3} = \frac{3(1 + 4x)}{3}.$$

Again, since $\frac{3}{3} = 1$, we have

$$\frac{3 + 12x}{3} = \frac{3(1 + 4x)}{3}$$

$$= \frac{3}{3}(1 + 4x)$$

$$= 1(1 + 4x)$$
$$= 1 + 4x.$$

It is important to remember that you can divide out only *factors* from a rational expression. For example, it is true that

$$\frac{15}{3} = \frac{3(5)}{3} = \frac{3}{3}(5) = 1(5) = 5.$$

However, clearly we cannot write $\frac{16}{3} = \frac{3+13}{3}$, and then write $\frac{3+13}{3}$ as $\frac{3}{3} + 13 = 1 + 13 = 14$. In the second case 3 is an *addend*. You must never divide out addends from a rational expression.

☐ **EXAMPLE 19.4**   Reduce $\dfrac{x^2 - 1}{x + 1}$.

Solution   A common error is to try to divide the x^2 term by the x term. However, we cannot do this, because these terms are addends and not factors. Therefore, we must factor the numerator:

$$\frac{x^2 - 1}{x + 1} = \frac{(x + 1)(x - 1)}{x + 1}.$$

Now, we see that the binomial $x + 1$ is a factor of both the numerator and the denominator, and may be divided out:

$$\frac{x^2 - 1}{x + 1} = \frac{(x + 1)(x - 1)}{x + 1}$$
$$= \left(\frac{x + 1}{x + 1}\right)(x - 1)$$
$$= 1(x - 1)$$
$$= x - 1.$$

☐ **EXAMPLE 19.5**   Reduce $\dfrac{x^2 + 3x}{3x + 9}$.

Solution   First we factor each polynomial:

$$\frac{x^2 + 3x}{3x + 9} = \frac{x(x + 3)}{3(x + 3)}.$$

We have a common factor $x + 3$, so we reduce the rational expression by dividing out this common factor:

$$\frac{x^2 + 3x}{3x + 9} = \frac{x(x + 3)}{3(x + 3)}$$
$$= \frac{x}{3}\left(\frac{x + 3}{x + 3}\right)$$
$$= \frac{x}{3}(1)$$
$$= \frac{x}{3}.$$

When we are certain we are dividing out factors, and not addends, we may skip from the first step to the last. The missing steps can be indicated by crossing out the common factors to show they have been divided out. Thus, we may write

$$\frac{x^2 + 3x}{3x + 9} = \frac{x(\cancel{x + 3})}{3(\cancel{x + 3})}$$

$$= \frac{x}{3}.$$

A rational expression is reduced to **lowest terms** when all common factors have been divided out from the numerator and the denominator.

EXAMPLE 19.6   Reduce $\dfrac{x^2 + 3x + 2}{x^2 - x - 2}$.

Solution   Factoring each polynomial, and dividing out the common binomial factor,

$$\frac{x^2 + 3x + 2}{x^2 - x - 2} = \frac{(\cancel{x + 1})(x + 2)}{(\cancel{x + 1})(x - 2)}$$

$$= \frac{x + 2}{x - 2}.$$

The remaining binomials $x + 2$ and $x - 2$ have no common factors (remember that you cannot divide out addends). Therefore, the rational expression has been reduced to lowest terms.

Factors such as $x + 2$ and $2 + x$ are common factors because

$$2 + x = x + 2.$$

However, factors such as $x - 2$ and $2 - x$ are not equal and are not common factors. We may create a common factor in this case by factoring out -1 from one of the binomials. For example,

$$2 - x = -1(x - 2).$$

EXAMPLE 19.7   Reduce $\dfrac{1 - x}{x^2 - 2x + 1}$.

Solution   We factor the denominator:

$$\frac{1 - x}{x^2 - 2x + 1} = \frac{1 - x}{(x - 1)(x - 1)}.$$

Now, $1 - x$ and $x - 1$ are not common factors, but we can factor -1 from $1 - x$ in the numerator:

$$\frac{1 - x}{x^2 - 2x + 1} = \frac{(-1)(\cancel{x - 1})}{(x - 1)(\cancel{x - 1})}$$

$$= \frac{-1}{x - 1}.$$

☐ Exercise 19.1

Reduce:

1. $\dfrac{5x}{15}$

2. $\dfrac{6x}{9y}$

3. $\dfrac{4x^2}{16x}$

4. $\dfrac{10x^2}{2x^4}$

5. $\dfrac{3x + 6}{6}$

6. $\dfrac{5 - 10x}{5}$

7. $\dfrac{15x + 20x^2}{30x}$

8. $\dfrac{9x^2 - 12x}{6x}$

9. $\dfrac{x^2 - 25}{x - 5}$

10. $\dfrac{x^2 + 2x + 1}{x + 1}$

11. $\dfrac{x + 10}{x^2 - 100}$

12. $\dfrac{x - 2}{x^2 - 4x + 4}$

13. $\dfrac{3x + 9}{x^2 + 6x + 9}$

14. $\dfrac{4x - 4}{x^2 - 4x + 3}$

15. $\dfrac{x^2 + 4x}{x^2 - 16}$

16. $\dfrac{2x^2 - 4x}{x^2 - 4}$

17. $\dfrac{ax + ay}{2x + 2y}$

18. $\dfrac{x^2 + 2x}{2x + 4}$

19. $\dfrac{x^2 - 3x + 2}{x^2 - 5x + 4}$

20. $\dfrac{x^2 - x - 6}{x^2 + 5x + 6}$

21. $\dfrac{3 - x}{x^2 - 9}$

22. $\dfrac{x^2 - 4x + 3}{1 - x}$

23. $\dfrac{x^2 - 4x + 4}{4 - x^2}$

24. $\dfrac{3 - 3x^2}{3x - 3}$

☐ 19.2 ☐ Multiplication and Division

To multiply two rational numbers, we simply multiply the numerators and multiply the denominators. For example,

$$\frac{1}{2} \cdot \frac{1}{3} = \frac{1(1)}{2(3)} = \frac{1}{6}.$$

A product should be reduced when possible. For example,

$$\frac{2}{9} \cdot \frac{3}{4} = \frac{2(3)}{9(4)} = \frac{6}{36} = \frac{1}{6}.$$

When a product can be reduced, we can divide out common factors before performing the multiplications. This approach simplifies both the multiplication and the reducing. The preceding example can be done in this way:

$$\frac{2}{9} \cdot \frac{3}{4} = \frac{1(\cancel{2})}{\cancel{3}(3)} \cdot \frac{1(\cancel{3})}{\cancel{2}(2)} = \frac{1(1)}{3(2)} = \frac{1}{6}.$$

Multiplication of rational expressions is exactly the same as multiplication of rational numbers. We multiply the two polynomials in the numerators and the two polynomials in the denominators.

☐ EXAMPLE 19.8   Multiply $\dfrac{z}{2} \cdot \dfrac{yz}{3x}$.

Solution   We write the products of the two numerators and of the two denominators:

$$\frac{z}{2} \cdot \frac{yz}{3x} = \frac{z(yz)}{2(3x)}$$

$$= \frac{yz^2}{6x}.$$

EXAMPLE 19.9   Multiply $\dfrac{3x^2}{4y^2} \cdot \dfrac{2y^3}{15x^4}$.

Solution   We may divide out any common factors in the numerators and the denominators:

$$\frac{3x^2}{4y^2} \cdot \frac{2y^3}{15x^2} = \frac{\cancel{3}\cancel{x^2}}{\cancel{2}(2)(\cancel{y^2})} \cdot \frac{\cancel{2}(\cancel{y^2})(y)}{\cancel{3}(5)(\cancel{x^2})(x^2)}$$

$$= \frac{y}{2(5x^2)}$$

$$= \frac{y}{10x^2}.$$

Now, suppose we have a multiplication of rational expressions in which some or all of the polynomials in the expressions can be factored. We will always factor each polynomial that can be factored, and reduce by dividing out common factors in the numerators and denominators.

EXAMPLE 19.10   Multiply $\dfrac{5x - 15}{x} \cdot \dfrac{x^2}{x^2 - 9}$.

Solution   First, we factor each polynomial:

$$\frac{5x - 15}{x} \cdot \frac{x^2}{x^2 - 9} = \frac{5(x - 3)}{x} \cdot \frac{(x)(x)}{(x + 3)(x - 3)}$$

The common factors are $x - 3$ and x. We divide out these factors and multiply:

$$\frac{5x - 15}{x} \cdot \frac{x^2}{x^2 - 9} = \frac{5(\cancel{x - 3})}{\cancel{x}} \cdot \frac{(\cancel{x})(x)}{(x + 3)(\cancel{x - 3})}$$

$$= \frac{5}{1} \cdot \frac{x}{x + 3}$$

$$= \frac{5x}{x + 3}.$$

The resulting rational expression is the product, reduced by dividing out all common factors. Observe that we cannot divide out the remaining x because it is an addend in the denominator. Remember that you may only divide out *factors*.

EXAMPLE 19.11   Multiply $\dfrac{2x + 4}{x^2 + 2x + 1} \cdot \dfrac{x^2 + x}{2x}$.

Solution   First, we factor each polynomial:

$$\frac{2x + 4}{x^2 + 2x + 1} \cdot \frac{x^2 + x}{2x} = \frac{2(x + 2)}{(x + 1)(x + 1)} \cdot \frac{x(x + 1)}{2x}.$$

We have the common factors 2, x, and $x + 1$. We divide out these factors to reduce the resulting expression:

$$\frac{2x + 4}{x^2 + 2x + 1} \cdot \frac{x^2 + x}{2x} = \frac{\cancel{2}(x + 2)}{\cancel{(x + 1)}(x + 1)} \cdot \frac{\cancel{x}\cancel{(x + 1)}}{\cancel{2x}}$$

$$= \frac{x + 2}{x + 1}.$$

EXAMPLE 19.12 Multiply $\dfrac{x^2 - 2x - 3}{x^2 - 4} \cdot \dfrac{x^2 + 4x + 4}{x^2 + x - 12}$.

Solution   A common error is to divide out parts that look alike, such as the x^2 terms or the constant terms. However, these parts are addends and not factors. Before we can divide out any common factors, we must factor each polynomial:

$$\frac{x^2 - 2x - 3}{x^2 - 4} \cdot \frac{x^2 + 4x + 4}{x^2 + x - 12} = \frac{(x + 1)(x - 3)}{(x + 2)(x - 2)} \cdot \frac{(x + 2)(x + 2)}{(x + 4)(x - 3)}.$$

Now we can divide out the common factors, and write the product in factored form:

$$\frac{x^2 - 2x - 3}{x^2 - 4} \cdot \frac{x^2 + 4x + 4}{x^2 + x - 12} = \frac{(x + 1)\cancel{(x - 3)}}{\cancel{(x + 2)}(x - 2)} \cdot \frac{\cancel{(x + 2)}(x + 2)}{(x + 4)\cancel{(x - 3)}}$$

$$= \frac{(x + 1)(x + 2)}{(x - 2)(x + 4)}.$$

The resulting expression cannot be reduced further.

The method of factoring -1 from a binomial, used in Section 19.1, can be used in multiplying rational expressions.

EXAMPLE 19.13 Multiply $\dfrac{9 - x^2}{3x} \cdot \dfrac{3x - 6}{x^2 - 5x + 6}$.

Solution   When we factor, we find we have factors $3 - x$ and $x - 3$, which are not common factors. We factor out -1 to obtain $3 - x = (-1)(x - 3)$:

$$\frac{9 - x^2}{3x} \cdot \frac{3x - 6}{x^2 - 5x + 6} = \frac{(3 - x)(3 + x)}{3x} \cdot \frac{3(x - 2)}{(x - 2)(x - 3)}$$

$$= \frac{(-1)\cancel{(x - 3)}(3 + x)}{\cancel{3}x} \cdot \frac{\cancel{3}\cancel{(x - 2)}}{\cancel{(x - 2)}\cancel{(x - 3)}}$$

$$= \frac{(-1)(3 + x)}{x}.$$

This result may be written in many forms, such as

$$\frac{-(x + 3)}{x}$$

or

$$-\frac{x + 3}{x}$$

or

$$\frac{-x-3}{x}.$$

Recall that, to divide two rational numbers, we invert the second number and multiply. For example,

$$\frac{1}{2} \div \frac{3}{4} = \frac{1}{2} \cdot \frac{4}{3} = \frac{1}{\cancel{2}} \cdot \frac{\cancel{2}(2)}{3} = \frac{2}{3}.$$

To see why "invert and multiply" works, we write the division as a fraction in which the numerator and denominator are each rational numbers:

$$\frac{1}{2} \div \frac{3}{4} = \frac{\dfrac{1}{2}}{\dfrac{3}{4}}.$$

We multiply the numerator and denominator of the fraction by $\frac{4}{3}$. Then we have the denominator 1:

$$\frac{1}{2} \div \frac{3}{4} = \frac{\dfrac{1}{2}}{\dfrac{3}{4}} = \frac{\dfrac{1}{2} \cdot \dfrac{4}{3}}{\dfrac{3}{4} \cdot \dfrac{4}{3}} = \frac{\dfrac{1}{2} \cdot \dfrac{4}{3}}{1} = \frac{1}{2} \cdot \frac{4}{3}.$$

For any rational number $\frac{a}{b}$, with $a \neq 0$ and $b \neq 0$, the rational number $\frac{b}{a}$ such that $\frac{a}{b} \cdot \frac{b}{a} = 1$ is called the **reciprocal** or **multiplicative inverse** of $\frac{a}{b}$. Thus, division by a rational number is the same as multiplication by its reciprocal, or multiplicative inverse.

The same method is used to divide rational expressions. We multiply by the reciprocal of the divisor.

EXAMPLE 19.14   Divide $\dfrac{4x^2z}{y^3} \div \dfrac{2x}{yz}$.

Solution   We multiply by the reciprocal of the divisor, which is the second rational expression:

$$\frac{4x^2z}{y^3} \div \frac{2x}{yz} = \frac{4x^2z}{y^3} \cdot \frac{yz}{2x}$$

$$= \frac{\cancel{2}(2)(\cancel{x})(x)(z)}{(\cancel{y})(y^2)} \cdot \frac{\cancel{yz}}{\cancel{2x}}$$

$$= \frac{(2xz)(z)}{y^2}$$

$$= \frac{2xz^2}{y^2}.$$

EXAMPLE 19.15   Divide $\dfrac{3x-6}{x^2+x} \div \dfrac{x^2-4}{x^2+2x+1}$.

Solution   First, we write the division as a multiplication by the reciprocal of the second rational expression:

$$\frac{3x - 6}{x^2 + x} \div \frac{x^2 - 4}{x^2 + 2x + 1} = \frac{3x - 6}{x^2 + x} \cdot \frac{x^2 + 2x + 1}{x^2 - 4}.$$

To multiply, we factor the polynomials and divide out common factors, as in the preceding examples:

$$\frac{3x - 6}{x^2 + x} \cdot \frac{x^2 + 2x + 1}{x^2 - 4} = \frac{3\cancel{(x - 2)}}{x\cancel{(x + 1)}} \cdot \frac{\cancel{(x + 1)}(x + 1)}{(x + 2)\cancel{(x - 2)}}$$

$$= \frac{3(x + 1)}{x(x + 2)}.$$

EXAMPLE 19.16   Divide $\dfrac{x}{x + y} \div \dfrac{x + y}{y}$.

Solution   In this kind of example, there is a temptation to divide out the factor $x + y$. However, the example is a division. When it is rewritten as a multiplication by the reciprocal of the second rational expression, both factors $x + y$ will be in the denominators and so cannot be divided out. The solution is

$$\frac{x}{x + y} \div \frac{x + y}{y} = \frac{x}{x + y} \cdot \frac{y}{x + y}$$

$$= \frac{xy}{(x + y)(x + y)}.$$

This expression may also be written in the form

$$\frac{xy}{(x + y)^2}.$$

The expression cannot be reduced.

Exercise 19.2

Multiply or divide as indicated:

1. $\dfrac{xy}{3} \cdot \dfrac{2y}{z}$

2. $\dfrac{yz}{4x} \cdot \dfrac{y}{2x}$

3. $\dfrac{9x^3}{2y^2} \cdot \dfrac{4y}{3x^2}$

4. $\dfrac{5x^3y^3}{z^2} \cdot \dfrac{yz^2}{10x^4y^2}$

5. $\dfrac{6x}{5y^2} \div \dfrac{3x^2}{10y}$

6. $\dfrac{2x^4y^2}{z^2} \div \dfrac{xy^2}{2z^2}$

7. $\dfrac{2x}{x + 2} \cdot \dfrac{x + 2}{4}$

8. $\dfrac{3x - 6}{3} \cdot \dfrac{3x}{x - 2}$

9. $\dfrac{x^2 - 3x}{3x} \cdot \dfrac{3}{x - 3}$

10. $\dfrac{5xy + 10x}{2xy + 4x} \cdot \dfrac{2y + 4}{5y + 5}$

11. $\dfrac{x+3}{6x} \div \dfrac{x+3}{2x}$

12. $\dfrac{x}{3x-3} \div \dfrac{1}{x-1}$

13. $\dfrac{x}{x-2} \div \dfrac{x-2}{2}$

14. $\dfrac{2x}{xy+y} \div \dfrac{2x+2}{y}$

15. $\dfrac{x^2+x}{1-x^2} \cdot \dfrac{x-1}{1+x}$

16. $\dfrac{y}{y-x} \div \dfrac{x+y}{x^2-y^2}$

17. $\dfrac{x^2-x}{x^2+3x+2} \cdot \dfrac{4x+4}{x^2-3x+2}$

18. $\dfrac{2-6x}{3x^2-5x-2} \cdot \dfrac{2x^2-3x-2}{4-12x}$

19. $\dfrac{3x-6}{x^2+5x+6} \div \dfrac{x^2-5x+6}{x^2+2x}$

20. $\dfrac{x-2x^2}{x-x^2} \div \dfrac{2x^2+x-1}{x^2+x-2}$

21. $\dfrac{2x^2+3x-9}{3x^2+4x-4} \cdot \dfrac{3x^2-8x+4}{2x^2-9x+9}$

22. $\dfrac{9x^2-3x-2}{3x^2+7x-6} \div \dfrac{9x^2-1}{3x^2+8x-3}$

23. $\dfrac{4-x^2}{2x+4} \cdot \dfrac{4x+4}{x^2-x-2}$

24. $\dfrac{3x-9}{y^2-x^2} \div \dfrac{6-2x}{x^2-2xy+y^2}$

19.3 Addition and Subtraction

Recall from arithmetic that, in order to add or subtract fractions, we must use a **common denominator.** In multiplying fractions, and similarly in multiplying rational expressions, we simply multiply the numerators and multiply the denominators. It is a common error in arithmetic and in algebra to try to use a similar process to add fractions or rational expressions.

Suppose we want to add $\frac{1}{2} + \frac{1}{2}$. Were we simply to add the numerators and the denominators, we would get $\frac{2}{4}$, or $\frac{1}{2}$. This answer clearly is not correct.

Using a common denominator, we write

$$\frac{1}{2} + \frac{1}{2} = \frac{1+1}{2} = \frac{2}{2} = 1,$$

which is the correct sum. The sum of fractions is the sum of the numerators written over a common denominator. We add rational expressions in a similar way.

EXAMPLE 19.17 Add $\dfrac{x-1}{x-3} + \dfrac{x-2}{x-3}$.

Solution

There is already a common denominator, $x-3$, so we write the sum of the numerators over this denominator:

$$\frac{x-1}{x-3} + \frac{x-2}{x-3} = \frac{(x-1)+(x-2)}{x-3}$$

$$= \frac{x-1+x-2}{x-3}$$

$$= \frac{2x-3}{x-3}.$$

This rational expression cannot be reduced. Remember that you cannot divide out addends such as the x terms or the constant terms.

To subtract fractions or rational expressions, we follow the same procedure. The difference of the numerators is written over a common denominator. However, when we are subtracting rational expressions, we must be careful to use parentheses in the numerator.

EXAMPLE 19.18   Subtract $\dfrac{x-1}{x-3} - \dfrac{x-2}{x-3}$.

Solution   We write the difference of the numerators over the common denominator, being careful to use parentheses to indicate the separate numerators:

$$\frac{x-1}{x-3} - \frac{x-2}{x-3} = \frac{(x-1)-(x-2)}{x-3}.$$

Now, we remove the parentheses, being careful to apply the minus sign to both terms of the second polynomial:

$$\frac{(x-1)-(x-2)}{x-3} = \frac{x-1-x+2}{x-3}$$

$$= \frac{1}{x-3}.$$

Observe that the parentheses remind you to apply the minus sign to all terms of the second polynomial in the numerator. You should always use the parentheses to avoid errors in signs often made in subtracting rational expressions.

EXAMPLE 19.19   Subtract $\dfrac{x^2 + 2x + 1}{x+2} - \dfrac{1 - 2x - x^2}{x+2}$.

Solution   We write the difference of the numerators over the common denominator, using parentheses, and then remove the parentheses:

$$\frac{x^2 + 2x + 1}{x+2} - \frac{1 - 2x - x^2}{x+2} = \frac{(x^2 + 2x + 1) - (1 - 2x - x^2)}{x+2}$$

$$= \frac{x^2 + 2x + 1 - 1 + 2x + x^2}{x+2}$$

$$= \frac{2x^2 + 4x}{x+2}.$$

This expression can be reduced by factoring the polynomial in the numerator:

$$\frac{2x^2 + 4x}{x+2} = \frac{2x(x+2)}{x+2}$$

$$= 2x.$$

We factor and reduce the resulting rational expression whenever possible.

To add or subtract rational expressions whose denominators are not the same, we find the **least common denominator,** or LCD. For example, recall that to add two rational numbers such as $\frac{1}{2}$ and $\frac{1}{3}$, we write

$$\frac{1}{2} + \frac{1}{3} = \frac{1(3)}{2(3)} + \frac{1(2)}{3(2)} = \frac{3}{6} + \frac{2}{6} = \frac{3+2}{6} = \frac{5}{6}.$$

The LCD in this example is 6.

The LCD of two rational numbers is the **least common multiple,** LCM, of their denominators. The LCM of two positive integers is the smallest positive integer divisible by each. For example, the LCM of 2 and 3 is 6, because 6 is divisible by 2 and by 3, and there is no smaller positive integer divisible by both 2 and 3.

If the given numbers have common factors, then the LCM is smaller than the product of the given numbers. For example, the LCM of 6 and 4 is 12, not 24. It is true that 24 is divisible by both 6 and 4, but 24 is not the smallest positive integer divisible by both 6 and 4. The smallest positive integer divisible by both 6 and 4 is 12.

To find the LCM of two positive integers, we factor each number completely:

$$6 = (2)(3)$$

and

$$4 = (2)(2)$$

Then, we take each factor the greatest number of times it appears in either of the numbers. The factor 2 appears once in 6 and twice in 4. The greatest number of times 2 appears is twice. Therefore, the LCM has two factors of 2. The factor 3 appears once in 6 and no times in 4. The greatest number of times 3 appears is once. Therefore, the LCM has one factor of 3. Thus, the LCM of 6 and 4 is

$$(2)(2)(3) = 12.$$

It is possible for the LCM of two or more positive integers to be one of the integers. For example, the factor 2 appears once in 2 and twice in 4, so the LCM of 2 and 4 is 4.

The same method is used to find the LCM of two polynomials. If two polynomials have no common factors, their LCM is their product. If two polynomials do have common factors, we take each factor the greatest number of times it appears in either of the polynomials.

We use the LCM of two polynomials to add or subtract two rational expressions whose denominators are not the same. The LCD of rational expressions is the LCM of the polynomials in their denominators.

EXAMPLE 19.20   Add $\dfrac{4}{x} + \dfrac{3}{x^2}$.

Solution   The factor x appears once in x and twice in x^2, so the LCM of x and x^2 is x^2. Therefore, the LCD of the rational expressions is x^2. We write the first rational expression with the denominator x^2 by multiplying the numerator and the denominator by x:

$$\frac{4}{x} = \frac{4(x)}{x(x)} = \frac{4x}{x^2}.$$

The second rational expression already has the denominator x^2. Therefore,

$$\frac{4}{x} + \frac{3}{x^2} = \frac{4x}{x^2} + \frac{3}{x^2}$$

$$= \frac{4x + 3}{x^2}.$$

The numerator and the denominator of the resulting rational expression have no common factors, so this result cannot be reduced.

EXAMPLE 19.21   Add $\dfrac{x - 1}{x} + \dfrac{x + 2}{x + 1}$.

Solution   Since x and $x + 1$ have no common factors, the LCM of x and $x + 1$ is $x(x + 1)$. Therefore, the LCD of the rational expressions is $x(x + 1)$. We must write each expression with this denominator. For the first expression, we multiply the numerator and the denominator by $x + 1$:

$$\frac{x - 1}{x} = \frac{(x - 1)(x + 1)}{x(x + 1)}.$$

For the second expression, we multiply the numerator and the denominator by x:

$$\frac{x + 2}{x + 1} = \frac{x(x + 2)}{x(x + 1)}.$$

Thus, we have

$$\frac{x - 1}{x} + \frac{x + 2}{x + 1} = \frac{(x - 1)(x + 1)}{x(x + 1)} + \frac{x(x + 2)}{x(x + 1)}.$$

Now, writing the sum of the numerators over the common denominator,

$$\frac{(x - 1)(x + 1)}{x(x + 1)} + \frac{x(x + 2)}{x(x + 1)} = \frac{(x - 1)(x + 1) + x(x + 2)}{x(x + 1)}.$$

Finally, we may multiply, remove parentheses, and simplify the numerator:

$$\frac{(x - 1)(x + 1) + x(x + 2)}{x(x + 1)} = \frac{(x^2 - 1) + (x^2 + 2x)}{x(x + 1)}$$

$$= \frac{x^2 - 1 + x^2 + 2x}{x(x + 1)}$$

$$= \frac{2x^2 + 2x - 1}{x(x + 1)}.$$

The simplified numerator cannot be factored, and so the resulting rational expression cannot be reduced. Again, remember that you cannot divide out addends, so you cannot reduce a rational expression unless the numerator and denominator have a common factor.

To subtract rational expressions whose denominators are not already the same, we follow essentially the same process. As before, we must remember to use parentheses carefully to be sure the signs of the second polynomial in the numerator are correct.

EXAMPLE 19.22 Subtract $\dfrac{x + 2}{x + 1} - \dfrac{x - 1}{x - 2}$.

Solution   The denominators have no common factors; therefore, the LCD is $(x + 1)(x - 2)$. We write

$$\frac{x + 2}{x + 1} = \frac{(x + 2)(x - 2)}{(x + 1)(x - 2)},$$

and

$$\frac{x - 1}{x - 2} = \frac{(x - 1)(x + 1)}{(x - 2)(x + 1)}.$$

Now, we write the difference of the numerators over the common denominator:

$$\frac{x + 2}{x + 1} - \frac{x - 1}{x - 2} = \frac{(x + 2)(x - 2)}{(x + 1)(x - 2)} - \frac{(x - 1)(x + 1)}{(x - 2)(x + 1)}$$

$$= \frac{(x + 2)(x - 2) - (x - 1)(x + 1)}{(x + 1)(x - 2)}.$$

We simplify the numerator, taking care in removing the parentheses:

$$\frac{(x + 2)(x - 2) - (x - 1)(x + 1)}{(x + 1)(x - 2)} = \frac{(x^2 - 4) - (x^2 - 1)}{(x + 1)(x - 2)}$$

$$= \frac{x^2 - 4 - x^2 + 1}{(x + 1)(x - 2)}$$

$$= \frac{-3}{(x + 1)(x - 2)}.$$

It may be necessary to factor the denominators of the algebraic expressions to find the LCD. When we factor the denominators, we can find any common factors.

EXAMPLE 19.23 Subtract $\dfrac{1}{x^2 + 5x + 6} - \dfrac{2}{x^2 + 6x + 8}$.

Solution   First, we find the LCD. Factoring,

$$x^2 + 5x + 6 = (x + 2)(x + 3)$$
$$x^2 + 6x + 8 = (x + 2)(x + 4).$$

The common factor $x + 2$ appears just once in each denominator. Therefore, the LCD is $(x + 2)(x + 3)(x + 4)$. Now, we write each rational expression with the LCD:

$$\frac{1}{x^2 + 5x + 6} = \frac{1}{(x + 2)(x + 3)} = \frac{1(x + 4)}{(x + 2)(x + 3)(x + 4)}.$$

and

$$\frac{2}{x^2 + 6x + 8} = \frac{2}{(x + 2)(x + 4)} = \frac{2(x + 3)}{(x + 2)(x + 3)(x + 4)}.$$

Writing the numerators over the common denominator and simplifying the resulting numerator,

$$\frac{1(x+4)}{(x+2)(x+3)(x+4)} - \frac{2(x+3)}{(x+2)(x+3)(x+4)} = \frac{1(x+4) - 2(x+3)}{(x+2)(x+3)(x+4)}$$

$$= \frac{x+4-2x-6}{(x+2)(x+3)(x+4)}$$

$$= \frac{-x-2}{(x+2)(x+3)(x+4)}.$$

In this example, we have a rational expression that can be reduced. We factor out -1 from the numerator, and divide out the common factor $x+2$:

$$\frac{-x-2}{(x+2)(x+3)(x+4)} = \frac{(-1)\cancel{(x+2)}}{\cancel{(x+2)}(x+3)(x+4)}$$

$$= \frac{-1}{(x+3)(x+4)}.$$

In a special case of adding or subtracting rational expressions, we can find the LCD by factoring out -1 from one of the denominators.

EXAMPLE 19.24 Add $\dfrac{y}{x-y} + \dfrac{x}{y-x}$.

Solution   We use $y - x = (-1)(x - y)$. Then, we use $x - y$ as the LCD:

$$\frac{y}{x-y} + \frac{x}{y-x} = \frac{y}{x-y} + \frac{x}{(-1)(x-y)}$$

$$= \frac{y}{x-y} - \frac{x}{x-y}$$

$$= \frac{y-x}{x-y}.$$

Again using $y - x = (-1)(x - y)$, this rational expression can be reduced:

$$\frac{y-x}{x-y} = \frac{(-1)\cancel{(x-y)}}{\cancel{x-y}}$$

$$= -1.$$

Observe that you must be careful not to add the numerators of the original rational expressions without constructing a common denominator. Also, you must be sure not to divide out addends at any step of the solution.

Exercise 19.3

Add or subtract as indicated:

1. $\dfrac{x}{x+3} + \dfrac{2x}{x+3}$

2. $\dfrac{x-1}{x+4} + \dfrac{x-3}{x+4}$

3. $\dfrac{3x}{x+1} - \dfrac{x}{x+1}$

4. $\dfrac{x-2}{x+2} - \dfrac{x-4}{x+2}$

5. $\dfrac{x^2+x+1}{x+1} - \dfrac{x+2}{x+1}$

6. $\dfrac{2x^2-2x+3}{x-1} - \dfrac{3+x-x^2}{x-1}$

7. $\dfrac{2}{x^2} - \dfrac{3}{x}$

8. $\dfrac{1}{x} + \dfrac{4}{x^3}$

9. $\dfrac{x+3}{x-2} + \dfrac{x-1}{x}$

10. $\dfrac{x-2}{x} - \dfrac{x+1}{x+4}$

11. $\dfrac{x+2}{x-1} + \dfrac{x-3}{x+1}$

12. $\dfrac{x+1}{x+2} + \dfrac{x-1}{x-2}$

13. $\dfrac{2x+1}{x-2} - \dfrac{x-1}{x+1}$

14. $\dfrac{x+4}{x+2} - \dfrac{x-2}{x-3}$

15. $\dfrac{-2}{x^2-2x-3} + \dfrac{3}{x^2-9}$

16. $\dfrac{1}{x^2-5x+6} + \dfrac{-2}{x^2-6x+8}$

17. $\dfrac{2}{x^2+2x} - \dfrac{1}{x^2+x}$

18. $\dfrac{8}{x^2+8x+15} - \dfrac{4}{x^2+7x+12}$

19. $\dfrac{3}{x-3} - \dfrac{x}{3-x}$

20. $\dfrac{y}{y-x} + \dfrac{x}{x-y}$

21. $\dfrac{4x}{16-x^2} + \dfrac{x}{x-4}$

22. $\dfrac{x}{x^2-1} + \dfrac{1}{1-x}$

23. $\dfrac{y^2}{y^2-x^2} + \dfrac{y}{x-y}$

24. $\dfrac{4}{x^2-4x+4} - \dfrac{4}{4-x^2}$

Self-test

1. Reduce $\dfrac{2x^2 - 6x}{x^2 - x - 6}$

1. _____

Perform the operation as indicated:

2. $\dfrac{x - 3}{x^2 + x} \div \dfrac{x^2 - 9}{x + 1}$

2. _____

3. $\dfrac{x^2 - 4x + 4}{x + 2} \cdot \dfrac{x + 1}{x^2 - x - 2}$

3. _____

4. $\dfrac{x - 4}{x^2 - 4} - \dfrac{2}{2 - x}$

4. _____

5. $\dfrac{x - 1}{x - 3} - \dfrac{x - 3}{x - 1}$

5. _____

20 Equations with Rational Expressions

INTRODUCTION

In the last several units, you have used a number of tools of algebra to work with expressions that are not linear expressions. In particular, in the preceding unit you used the least common denominator to add and subtract rational expressions. In this unit you will learn how to use the least common denominator to solve equations involving rational expressions. Then, you will use equations involving rational expressions to solve a variation of the familiar distance-rate-time problem, and also in a new application.

OBJECTIVES

When you have finished this unit you should be able to:

1. Solve equations involving rational expressions.
2. Use equations involving rational expressions to solve uniform motion problems and work problems.

20.1 Solving Equations with Rational Expressions

In Section 7.3, we solved equations with constant denominators. To solve such an equation, we multiplied both sides of the equation by the LCD. As in Section 19.3, to find the LCD we find the LCM of the denominators.

EXAMPLE 20.1   Solve $\dfrac{x}{6} + \dfrac{5}{12} = \dfrac{3x}{8}$.

Solution   Using the method of Section 19.3, we factor each denominator:

$$6 = (2)(3),$$
$$12 = (2)(2)(3),$$
$$8 = (2)(2)(2).$$

The factor 3 appears at most once in any denominator, and the factor 2 appears at most three times. Therefore, the LCM of the denominators is

$$(2)(2)(2)(3) = 24.$$

The LCD is 24. Multiplying both sides of the equation by 24,

$$\frac{x}{6} + \frac{5}{12} = \frac{3x}{8}$$

$$24\left(\frac{x}{6} + \frac{5}{12}\right) = 24\left(\frac{3x}{8}\right).$$

In Section 7.3, we observed that multiplying both sides of the equation by the LCD and then using the distributive property is the same as multiplying each term of the equation by the LCD:

$$24\left(\frac{x}{6}\right) + 24\left(\frac{5}{12}\right) = 24\left(\frac{3x}{8}\right).$$

Now, we divide out the common factors in each term and solve the resulting equation:

$$4x + 2(5) = 3(3x)$$
$$4x + 10 = 9x$$
$$10 = 5x$$
$$2 = x.$$

The solution is 2. To check, we substitute 2 for x in the original equation:

$$\frac{x}{6} + \frac{5}{12} = \frac{3x}{8}$$
$$\frac{2}{6} + \frac{5}{12} \stackrel{?}{=} \frac{3(2)}{8}$$
$$\frac{4}{12} + \frac{5}{12} \stackrel{?}{=} \frac{3}{4}$$
$$\frac{9}{12} \stackrel{?}{=} \frac{3}{4}$$
$$\frac{3}{4} = \frac{3}{4}.$$

We may use the same procedure for equations in which the denominators are not constants. However, we cannot multiply both sides of an equation by zero. Therefore, when the LCD contains a variable, we must be sure that the variable does not take on a value that will make the LCD zero.

EXAMPLE 20.2   Solve $\dfrac{1}{x} + 3 = \dfrac{2}{x}$.

Solution   The LCD is x. We will assume x is not zero, because the denominator of a rational expression cannot be zero. Multiplying both sides of the equation by x, we have

$$\frac{1}{x} + 3 = \frac{2}{x}$$
$$x\left(\frac{1}{x} + 3\right) = x\left(\frac{2}{x}\right)$$
$$x\left(\frac{1}{x}\right) + x(3) = x\left(\frac{2}{x}\right).$$

Observe that multiplying both sides of the equation by the LCD is the same as multiplying each term by the LCD. We will usually skip the middle step and begin by multiplying each term by the LCD.
Dividing out common factors in each term,

$$\cancel{x}\left(\frac{1}{\cancel{x}}\right) + x(3) = \cancel{x}\left(\frac{2}{\cancel{x}}\right)$$
$$1 + 3x = 2.$$

Now, we solve the resulting linear equation:

$$1 + 3x = 2$$

$$3x = 1$$

$$x = \frac{1}{3}.$$

Since x is not zero, we have not multiplied by zero, and the solution is $\frac{1}{3}$. To check, we substitute $\frac{1}{3}$ for x in the original equation:

$$\frac{1}{x} + 3 = \frac{2}{x}$$

$$\frac{1}{\frac{1}{3}} + 3 \overset{?}{=} \frac{2}{\frac{1}{3}}$$

$$\frac{1}{1} \cdot \frac{3}{1} + 3 \overset{?}{=} \frac{2}{1} \cdot \frac{3}{1}$$

$$3 + 3 = 6.$$

Sometimes we will find an apparent solution that makes the LCD zero. Since we cannot multiply both sides of an equation by zero, such a number cannot actually be a solution. The number is an extraneous solution. We do not list extraneous solutions as solutions of an equation.

EXAMPLE 20.3   Solve $1 + \dfrac{10}{x - 5} = \dfrac{2x}{x - 5}$.

Solution   The LCD is $x - 5$:

$$1 + \frac{10}{x - 5} = \frac{2x}{x - 5}$$

$$(x - 5)(1) + (x - 5)\left(\frac{10}{x - 5}\right) = (x - 5)\left(\frac{2x}{x - 5}\right)$$

$$x - 5 + 10 = 2x$$

$$5 = x.$$

The apparent solution is 5. However, if $x = 5$, the LCD is zero. Therefore, 5 is an extraneous solution. Observe that if you try to check the extraneous solution, the denominators of the rational expressions are zero, and the rational expressions are undefined. Since there are no other apparent solutions, there is no solution to this equation.

As in Unit 19, we may factor -1 from a binomial to find the LCD.

EXAMPLE 20.4   Solve $\dfrac{1}{1 - x} - \dfrac{1}{2} = \dfrac{3}{x - 1}$.

Solution   Since $1 - x = (-1)(x - 1)$, we may write:

$$\frac{1}{1 - x} - \frac{1}{2} = \frac{3}{x - 1}$$

$$\frac{1}{(-1)(x-1)} - \frac{1}{2} = \frac{3}{x-1}$$

$$-\frac{1}{x-1} - \frac{1}{2} = \frac{3}{x-1}.$$

The LCD is $2(x-1)$:

$$-2(x-1)\left(\frac{1}{x-1}\right) - 2(x-1)\left(\frac{1}{2}\right) = 2(x-1)\left(\frac{3}{x-1}\right)$$

$$-2 - (x-1) = 6$$

$$-2 - x + 1 = 6$$

$$-1 - x = 6$$

$$-x = 7$$

$$x = -7.$$

The solution is -7. You should check this solution in the original equation.

When the denominators in an equation have no common factors, the LCD is the product of the denominators. When there are common factors, we take each factor the greatest number of times it appears in any one denominator.

EXAMPLE 20.5   Solve $\dfrac{1}{x} + \dfrac{1}{x+1} = \dfrac{11}{6x}$.

Solution   The factor x appears at most once in any one denominator. Thus, the LCD is $6x(x+1)$. We multiply each term by the LCD:

$$\frac{1}{x} + \frac{1}{x+1} = \frac{11}{6x}$$

$$6x(x+1)\left(\frac{1}{x}\right) + 6x(x+1)\left(\frac{1}{x+1}\right) = 6x(x+1)\left(\frac{11}{6x}\right).$$

Now, we divide out the common factors and solve the resulting equation:

$$6x(x+1)\left(\frac{1}{x}\right) + 6x(x+1)\left(\frac{1}{x+1}\right) = 6x(x+1)\left(\frac{11}{6x}\right)$$

$$6(x+1)(1) + 6x(1) = (x+1)(11)$$

$$6x + 6 + 6x = 11x + 11$$

$$12x + 6 = 11x + 11$$

$$x = 5.$$

The solution is 5. You should check this solution in the original equation.

An equation involving rational expressions may result in a quadratic equation. If the quadratic expression can be factored, we solve the equation using the methods of Unit 18.

EXAMPLE 20.6   Solve $\dfrac{2}{x} + \dfrac{x}{x+1} = \dfrac{5}{3}$.

Solution   The LCD is $3x(x + 1)$. We multiply each term by the LCD:

$$\frac{2}{x} + \frac{x}{x + 1} = \frac{5}{3}$$

$$3x(x + 1)\left(\frac{2}{x}\right) + 3x(x + 1)\left(\frac{x}{x + 1}\right) = 3x(x + 1)\left(\frac{5}{3}\right).$$

Dividing out common factors in each term,

$$3(x + 1)(2) + 3x(x) = x(x + 1)(5).$$

Now we may remove parentheses in each expression and simplify:

$$6(x + 1) + 3x(x) = 5x(x + 1)$$
$$6x + 6 + 3x^2 = 5x^2 + 5x.$$

The resulting equation is a quadratic equation. We can collect the terms on the right-hand side and solve by factoring:

$$0 = 2x^2 - x - 6$$
$$0 = (x - 2)(2x + 3)$$
$$x - 2 = 0 \text{ and } 2x + 3 = 0$$
$$x = 2 \text{ and } x = -\frac{3}{2}.$$

Both 2 and $-\frac{3}{2}$ are solutions. To check, we first substitute $x = 2$ in the original equation:

$$\frac{2}{x} + \frac{x}{x + 1} = \frac{5}{3}$$
$$\frac{2}{2} + \frac{2}{2 + 1} \stackrel{?}{=} \frac{5}{3}$$
$$1 + \frac{2}{3} = \frac{5}{3}.$$

Then we substitute $x = -\frac{3}{2}$:

$$\frac{2}{x} + \frac{x}{x + 1} = \frac{5}{3}$$

$$\frac{2}{-\frac{3}{2}} + \frac{-\frac{3}{2}}{-\frac{3}{2} + 1} \stackrel{?}{=} \frac{5}{3}$$

$$\frac{2}{-\frac{3}{2}} + \frac{-\frac{3}{2}}{-\frac{1}{2}} \stackrel{?}{=} \frac{5}{3}$$

$$\frac{2}{1}\left(-\frac{2}{3}\right) + \left(-\frac{3}{2}\right)\left(-\frac{2}{1}\right) \stackrel{?}{=} \frac{5}{3}$$

$$-\frac{4}{3} + 3 \overset{?}{=} \frac{5}{3}$$

$$-\frac{4}{3} + \frac{9}{3} = \frac{5}{3}.$$

As in Section 19.3, it may be necessary to factor the denominators in an equation to find common factors.

EXAMPLE 20.7   Solve $\dfrac{3x}{x^2 + 6x + 9} - \dfrac{6}{x + 3} = 0$.

Solution   Since $x^2 + 6x + 9 = (x + 3)(x + 3)$, the LCD is $(x + 3)(x + 3)$. We multiply each term by the LCD:

$$\frac{3x}{x^2 + 6x + 9} - \frac{6}{x + 3} = 0$$

$$(x + 3)(x + 3)\left(\frac{3x}{x^2 + 6x + 9}\right) - (x + 3)(x + 3)\left(\frac{6}{x + 3}\right) = (x + 3)(x + 3)(0).$$

$$3x - (x + 3)(6) = 0$$

$$3x - 6x - 18 = 0$$

$$-3x = 18$$

$$x = -6.$$

The solution is -6. To check,

$$\frac{3x}{x^2 + 6x + 9} - \frac{6}{x + 3} = 0$$

$$\frac{3(-6)}{(-6)^2 + 6(-6) + 9} - \frac{6}{-6 + 3} \overset{?}{=} 0$$

$$\frac{-18}{36 - 36 + 9} - \frac{6}{-3} \overset{?}{=} 0$$

$$\frac{-18}{9} - (-2) \overset{?}{=} 0$$

$$-2 + 2 = 0.$$

EXAMPLE 20.8   Solve $\dfrac{x}{x - 1} - \dfrac{4}{x + 2} = \dfrac{4}{x^2 + x - 2}$.

Solution   Since $x^2 + x - 2 = (x - 1)(x + 2)$, the LCD is $(x - 1)(x + 2)$:

$$\frac{x}{x - 1} - \frac{4}{x + 2} = \frac{4}{x^2 + x - 2}$$

$$(x - 1)(x + 2)\left(\frac{x}{x - 1}\right) - (x - 1)(x + 2)\left(\frac{4}{x + 2}\right) = (x - 1)(x + 2)\left(\frac{4}{x^2 + x - 2}\right)$$

$$(x + 2)(x) - (x - 1)(4) = 4$$

$$x^2 + 2x - 4x + 4 = 4$$

$$x^2 - 2x = 0.$$

We solve this quadratic equation by factoring out the common factor x:

$$x(x - 2) = 0$$
$$x = 0 \text{ and } x - 2 = 0$$
$$x = 0 \text{ and } x = 2.$$

The solutions are 0 and 2. The solution 0 is not the same as an LCD of 0; it does not result in an undefined rational expression. To check,

$$\frac{x}{x - 1} - \frac{4}{x + 2} = \frac{4}{x^2 + x - 2}$$

$$\frac{0}{0 - 1} - \frac{4}{0 + 2} \overset{?}{=} \frac{4}{0^2 + 0 - 2}$$

$$\frac{0}{-1} - \frac{4}{2} \overset{?}{=} \frac{4}{-2}$$

$$0 - 2 = -2.$$

You should check the solution 2.

Exercise 20.1

Solve and check:

1. $\dfrac{x}{2} + \dfrac{2}{9} = \dfrac{2x}{3}$

2. $\dfrac{5x}{6} - \dfrac{1}{12} = \dfrac{3x}{4} - \dfrac{1}{8}$

3. $\dfrac{3}{x} - 2 = \dfrac{5}{x}$

4. $\dfrac{1}{x} = 2 - \dfrac{2}{x}$

5. $\dfrac{4x}{2x + 3} = 1 - \dfrac{4}{2x + 3}$

6. $\dfrac{6}{x - 2} + 3 = \dfrac{x}{x - 2}$

7. $\dfrac{1}{x - 1} - 4 = \dfrac{x}{x - 1}$

8. $\dfrac{1}{2x - 1} = \dfrac{2x}{2x - 1} + 3$

9. $\dfrac{x}{x - 2} - \dfrac{2}{2 - x} = 3$

10. $\dfrac{1}{1 - 2x} + \dfrac{1}{2} = \dfrac{2x}{2x - 1}$

11. $\dfrac{2}{x} + \dfrac{1}{x - 1} = \dfrac{5}{2x}$

12. $\dfrac{3}{4x} + \dfrac{1}{x - 2} = \dfrac{3}{2x}$

13. $\dfrac{3}{x - 1} - \dfrac{3}{x} = \dfrac{4}{3x}$

14. $\dfrac{1}{x + 2} - \dfrac{1}{2x} = \dfrac{2}{3x}$

15. $\dfrac{12}{x} = \dfrac{2}{x - 3} + 1$

16. $\dfrac{8}{x + 3} + \dfrac{2}{x} = 2$

17. $\dfrac{8}{x + 2} - \dfrac{1}{x} = \dfrac{3}{2}$

18. $\dfrac{2x}{2x - 1} - \dfrac{3}{2x} = \dfrac{1}{2}$

19. $\dfrac{4x}{x^2 + 4x + 4} - \dfrac{1}{x + 2} = 0$

20. $\dfrac{5x}{x^2 - 9} - \dfrac{5}{x - 3} - \dfrac{3}{x + 3} = 0$

21. $\dfrac{2}{x - 2} - \dfrac{4}{x + 1} = \dfrac{2}{x^2 - x - 2}$

22. $\dfrac{1}{x - 2} + \dfrac{1}{x - 3} = \dfrac{1}{x^2 - 5x + 6}$

23. $\dfrac{x}{x - 2} - \dfrac{2}{x + 2} = \dfrac{5}{x^2 - 4}$

24. $\dfrac{x}{x - 1} - \dfrac{4}{x + 6} + \dfrac{2x}{x^2 + 5x - 6} = 0$

20.2 Applications

Recall the distance, rate, and time formula

$$D = RT,$$

used in Sections 2.4, 9.2, and 12.4. This formula can be solved for T to obtain

$$T = \frac{D}{R}.$$

We use this form of the formula to solve applied problems where a total time is given.

EXAMPLE 20.9 You walk to town on a route mostly uphill. Returning downhill, you walk twice as fast. The distance each way is 4 miles. If your total walking time is 3 hours, what is your average rate each way?

Solution As in Unit 8, we construct a chart for the problem. The headings are rate, time, and distance, where rate times time equals distance. If x represents your downhill rate, the rates are x and $2x$. The distance each way is 4. We use

$$T = \frac{D}{R}$$

to find the times, $\dfrac{4}{x}$ and $\dfrac{4}{2x}$:

	Rate	Time	Distance
Uphill	x	$\dfrac{4}{x}$	4
Downhill	$2x$	$\dfrac{4}{2x}$	4
Total		3	

Since the total time is 3 hours, we add the times each way to obtain the equation

$$\frac{4}{x} + \frac{4}{2x} = 3$$

or

$$\frac{4}{x} + \frac{2}{x} = 3.$$

Then, multiplying each term by x,

$$x\left(\frac{4}{x}\right) + x\left(\frac{2}{x}\right) = x(3)$$
$$4 + 2 = 3x$$
$$6 = 3x$$
$$2 = x.$$

Your average rate uphill is 2 miles per hour and your average rate downhill is twice that, or 4 miles per hour. To check, we use

$$\frac{D}{R} = T.$$

Your time walking uphill is

$$\frac{4}{2} = 2,$$

and your time walking downhill is

$$\frac{4}{4} = 1.$$

Therefore, the total time is $2 + 1 = 3$ hours.

EXAMPLE 20.10   It takes a boat a total of $3\frac{1}{2}$ hours to go to an island 18 nautical miles away and to return. Because of an outward tide, the boat returns at an average rate $\frac{3}{4}$ of its rate going out. What is its average rate each way?

Solution   We know that the rates are x and $\frac{3}{4}x$, and the distance each way is 18 nautical miles. The times are $\dfrac{18}{x}$ and $\dfrac{18}{\frac{3}{4}x}$:

	Rate	Time	Distance
Out	x	$\dfrac{18}{x}$	18
Return	$\dfrac{3}{4}x$	$\dfrac{18}{\frac{3}{4}x}$	18
Total		$\dfrac{7}{2}$	

We find that

$$\frac{18}{\frac{3}{4}x} = \frac{4}{3}\left(\frac{18}{x}\right) = \frac{24}{x}.$$

Therefore, the equation is

$$\frac{18}{x} + \frac{24}{x} = \frac{7}{2}$$

$$2x\left(\frac{18}{x}\right) + 2x\left(\frac{24}{x}\right) = 2x\left(\frac{7}{2}\right)$$

$$36 + 48 = 7x$$

$$84 = 7x$$

$$12 = x.$$

The average rate of the boat going out to the island is 12 knots (nautical miles per hour). Returning, the average rate is $\frac{3}{4}(12)$, or 9 knots. To check, the time to go to the island is

$$\frac{D}{R} = T$$

$$\frac{18}{12} = \frac{3}{2},$$

and the time to return is

$$\frac{18}{9} = 2.$$

Therefore, the total time is $\frac{3}{2} + 2 = \frac{7}{2} = 3\frac{1}{2}$ hours.

Work problems lead to equations similar to those encountered in distance, rate, and time problems. Work problems are based on the principle that the product of the rate at which work is done and the time spent on the work is equal to the amount of work done. For example, if a copier can copy 10 pages per minute, and works for 30 minutes, it can copy 300 pages.

Suppose one copier can copy 10 pages per minute and another can copy 15 pages per minute. It would take the first copier 30 minutes to copy 300 pages since $10(30) = 300$, and the second copier 20 minutes since $15(20) = 300$. Now, suppose both copiers are used. If the time using both copiers is x, the amount of work done by the first copier is

$$10x$$

and the amount of work done by the second copier is

$$15x.$$

The total amount of work is 300 pages, so

$$10x + 15x = 300$$

$$25x = 300$$

$$x = 12.$$

Using both copiers together, the 300 pages can be copied in 12 minutes.

The formula is

$$RT = A,$$

where R is the rate at which the work is done, T is the time for the work to be done, and A is the amount of work done. This formula can be solved for T to obtain

$$T = \frac{A}{R},$$

or for R to obtain

$$R = \frac{A}{T}.$$

A common type of work problem uses the third form.

EXAMPLE 20.11   One copier can copy a 120 page report in 6 minutes, while another takes 10 minutes. If both copiers are used, how long will it take to copy the report?

Solution   The first copier copies 120 pages in 6 minutes, or at a rate of $\frac{120}{6} = 20$ pages per minute. The second copier copies 120 pages in 10 minutes, or at a rate of $\frac{120}{10} = 12$ pages per minute. If both copiers working at the same time take x minutes, the rate using both is $\frac{120}{x}$. We put a third line on the chart to represent both copiers:

	Rate	Time	Amount
Copier 1	20	6	120
Copier 2	12	10	120
Both	$\frac{120}{x}$	x	120

Adding the rates,

$$20 + 12 = \frac{120}{x}$$

$$32 = \frac{120}{x}$$

$$32x = 120$$

$$x = \frac{120}{32}$$

$$x = 3\frac{3}{4}.$$

It takes $3\frac{3}{4}$ minutes to copy the report using both copiers. To check, we find the amount each copier does in $3\frac{3}{4}$ minutes. The first copier copies

$$20\left(\frac{15}{4}\right) = 75 \text{ pages.}$$

The second copier copies

$$12\left(\frac{15}{4}\right) = 45 \text{ pages.}$$

The total amount of pages copied in $3\frac{3}{4}$ minutes is then

$$75 + 45 = 120 \text{ pages.}$$

Often a work problem is stated in terms of the amount of time to do one total job. In this case we write the amount of work as 1, for one job.

EXAMPLE 20.12 It takes 9 hours for one machine to clean the carpeting in a house, and 6 hours for another to do the same job. How long will the job take if both machines are used?

Solution We make a chart using 1 for the amount:

	Rate	Time	Amount
Machine 1	$\frac{1}{9}$	9	1
Machine 2	$\frac{1}{6}$	6	1
Both	$\frac{1}{x}$	x	1

Then, adding the rates,

$$\frac{1}{9} + \frac{1}{6} = \frac{1}{x}$$

$$18x\left(\frac{1}{9}\right) + 18x\left(\frac{1}{6}\right) = 18x\left(\frac{1}{x}\right)$$

$$2x + 3x = 18$$

$$5x = 18$$

$$x = \frac{18}{5}$$

$$x = 3\frac{3}{5}.$$

It takes $3\frac{3}{5}$ hours using both machines. To check, in $3\frac{3}{5}$ hours the first machine does

$$\frac{1}{9}\left(\frac{18}{5}\right) = \frac{2}{5}$$

of the job, and the second machine does

$$\frac{1}{6}\left(\frac{18}{5}\right) = \frac{3}{5}$$

of the job, so the whole job is done.

EXAMPLE 20.13   Two tractors plow a field in $1\frac{1}{3}$ hours. If the first tractor could do the job alone in 2 hours, how long would it take the second tractor to do the job alone?

Solution   In this example, the time for the second tractor is x:

	Rate	Time	Amount
Tractor 1	$\dfrac{1}{2}$	2	1
Tractor 2	$\dfrac{1}{x}$	x	1
Both	$\dfrac{3}{4}$	$\dfrac{4}{3}$	1

Observe that the rate for both is the reciprocal of $1\frac{1}{3}$, that is, the reciprocal of $\frac{4}{3}$, or $\frac{3}{4}$. Then,

$$\frac{1}{2} + \frac{1}{x} = \frac{3}{4}$$
$$4x\left(\frac{1}{2}\right) + 4x\left(\frac{1}{x}\right) = 4x\left(\frac{3}{4}\right)$$
$$2x + 4 = 3x$$
$$4 = x$$
$$x = 4.$$

It would take the second tractor 4 hours. To check, in $1\frac{1}{3}$ hours, the first tractor does

$$\frac{1}{2}\left(\frac{4}{3}\right) = \frac{2}{3}$$

of the job, and the second tractor does

$$\frac{1}{4}\left(\frac{4}{3}\right) = \frac{1}{3}$$

of the job, so the whole job is done.

We must be careful not to follow the methods for solving applied problems blindly. It is easy to make up problems where the methods above do not make sense. With a little common sense, we can identify such problems.

EXAMPLE 20.14   Working with an old adding machine, it takes me 12 hours to do my tax returns. With my new calculator, I can do the job in 8 hours. How long will it take me using both machines?

Solution   It will take 8 hours. I can't use both machines at once, so the fastest way is to do the whole job with the calculator.

Exercise 20.2

1. A bus goes to town twice as fast as it returns. The town is 10 miles away. If the total time for the round trip is $1\frac{1}{2}$ hours, what is the average rate of the bus each way?

2. A boat sails 3 miles across a lake, and returns against the wind at an average rate $\frac{1}{4}$ as fast. If the total trip takes $2\frac{1}{2}$ hours, what is the average rate of the boat each way?

3. You walk for 2 miles, and then jog for 4 miles at an average rate twice as fast. Your total time is 1 hour. What are your average rate walking and your average rate jogging?

4. You bicycle 1 mile from your cabin to a lake, and then take a motor boat 8 miles across the lake. You bicycle at an average rate $\frac{3}{10}$ of the rate of the boat. If the total time for the trip is $\frac{2}{3}$ of an hour, what are your average rate bicycling and the average rate of the boat?

5. A copier can duplicate a 300 page document in 20 minutes, but it takes another copier 30 minutes. How long will it take if both copiers are used?

6. A new machine can cap 100 bottles in $3\frac{1}{3}$ minutes. It takes the old machine 5 minutes to cap 100 bottles. How long will it take to cap 100 bottles if both machines are used?

7. One boy scout can peel all the potatoes for the camp's dinner in 3 hours, and another takes 5 hours. If they do the job together, how long will it take?

8. Three farmers have a tractor that can plow a field in 12 hours, one that takes 8 hours, and one that takes 6 hours. If they use all three tractors, how long will it take to plow the field?

9. Using two machines, a road can be surfaced in 4 hours. One of the machines can do the job alone in 6 hours. How long would it take the other machine to do the job alone?

10. One newspaper carrier can cover a route in 2 hours. If the route is split with a second carrier, the two together take $1\frac{1}{4}$ hours. How long would it take the second carrier to cover the route alone?

11. One washing machine can wash a load in 30 minutes, and a second machine takes 40 minutes. How long will it take the two machines to wash one load?

12. You can walk to town in 3 hours, but your two friends take only $2\frac{1}{2}$ hours. How long will it take all three of you walking together?

Self-test

Solve and check:

1. $2 - \dfrac{x}{x+4} = \dfrac{4}{x+4}$

1. _____

2. $\dfrac{3x}{x-2} - \dfrac{5}{2} = \dfrac{2}{x}$

2. _____

3. $\dfrac{1}{3} + \dfrac{x}{3-x} = \dfrac{3}{x-3}$

3. _____

4. $\dfrac{2}{x-4} - \dfrac{1}{x+3} - \dfrac{12}{x^2-x-12} = 0$

4. _____

5. A boat goes 15 miles downstream, and returns upstream $\frac{1}{3}$ as fast. If the total trip takes $6\frac{2}{3}$ hours, what is the average rate of the boat each way?

5. _____

294

UNIT

Quadratic Equations

INTRODUCTION

Throughout most of this book you used linear expressions, in which the terms are at most first-degree. You learned how to solve linear equations and inequalities in one variable, how to use linear equations in some applications, and how to draw graphs of linear equations and inequalities in two variables. In the latter part of this book you have used quadratic expressions, where at least one term is a second-degree term involving the square of a variable. You learned a special method for solving quadratic equations, by factoring into linear factors over the integers. In this unit you will preview the next level of algebra. You will learn about square roots, general methods for solving quadratic equations, and how to draw the graph of one form of quadratic equation in two variables.

OBJECTIVES

When you have finished this unit you should be able to:

1. Reduce square roots of positive integers and positive rational numbers.
2. Solve equations of the form $ax^2 + c = 0$ when the equation has real number solutions, or state that the equation has no real number solutions.
3. Use the quadratic formula to solve quadratic equations when the equation has real number solutions, or state that the equation has no real number solutions.
4. Use the quadratic formula to solve geometric problems.
5. Find the y-intercept, x-intercepts, and vertex of a parabola, and draw its graph.

21.1 Square Roots

In Section 18.4, we had a brief introduction to square roots. We begin further discussion of square roots with the definition of a positive square root. Recall the symbol $\sqrt{a}$ (read "the square root of a").

Definition: $\sqrt{a} = x$ if $a = x^2$, where $a > 0$.

We repeat some examples from Section 18.4:

$$\sqrt{1} = 1 \text{ because } 1 = 1^2,$$
$$\sqrt{4} = 2 \text{ because } 4 = 2^2,$$
$$\sqrt{9} = 3 \text{ because } 9 = 3^2,$$
$$\sqrt{16} = 4 \text{ because } 16 = 4^2,$$
$$\sqrt{25} = 5 \text{ because } 25 = 5^2,$$

and so on. Numbers such as 1, 4, 9, 16, and 25 are called **perfect squares.** The positive square root of a perfect square is a positive integer. Also, $\sqrt{0} = 0$.

There are also negative square roots. For example,

$$-\sqrt{4} = -2.$$

Observe that

$$4 = (-2)^2$$

and therefore -2 is also a square root of 4. Any positive number a has both a positive square root $\sqrt{a}$ and a negative square root $-\sqrt{a}$.

Now, suppose a is negative. For example, consider $\sqrt{-4}$. We need a number x such that $-4 = x^2$. But if x is a positive number, then x^2 is positive. Also, if x is a negative number, then x^2 is a negative times a negative and again is positive. We do not know any number x such that $x^2 = -4$. If a is negative, we say that $\sqrt{a}$ is **not a real number.**

EXAMPLE 21.1   Find the value:

a. $\sqrt{121}$ b. $-\sqrt{81}$ c. $\sqrt{-100}$

Solutions

    a. $\sqrt{121} = 11$ because $121 = 11^2$.
    b. $-\sqrt{81} = -9$ because $81 = (-9)^2$.
    c. $\sqrt{-100}$ is not a real number.

We use two rules for square roots to find square roots of larger integers and to simplify square roots of integers and rational numbers. The proofs of these rules are given at the end of this unit.

Rules for Square Roots: If a and b are positive integers,

    1. $\sqrt{ab} = \sqrt{a}\sqrt{b}$

    2. $\sqrt{\dfrac{a}{b}} = \dfrac{\sqrt{a}}{\sqrt{b}}$

Rule 1 may be useful in finding square roots of larger integers.

EXAMPLE 21.2   Find the value of $\sqrt{400}$.

Solution

We may use Rule 1 to write

$$\begin{aligned}
\sqrt{400} &= \sqrt{(4)(100)} \\
&= \sqrt{4}\sqrt{100} \\
&= (2)(10) \\
&= 20.
\end{aligned}$$

Therefore,

$$\sqrt{400} = 20.$$

We may also use Rule 1 to simplify square roots of positive integers that are not perfect squares.

EXAMPLE 21.3   Simplify:

a. $\sqrt{50}$ b. $\sqrt{48}$ c. $\sqrt{69}$

Solutions

a. Fifty is not a perfect square. However, the perfect square 25 is a factor of 50. Using Rule 1,

$$\begin{aligned}
\sqrt{50} &= \sqrt{(25)(2)} \\
&= \sqrt{25}\sqrt{2} \\
&= 5\sqrt{2}.
\end{aligned}$$

b. Four is a factor of 48 and is a perfect square. Thus, we may write

$$\begin{aligned}
\sqrt{48} &= \sqrt{(4)(12)} \\
&= \sqrt{4}\sqrt{12} \\
&= 2\sqrt{12}.
\end{aligned}$$

However, 4 is also a factor of 12, and so the square root can be reduced further:

$$\begin{aligned}
2\sqrt{12} &= 2\sqrt{(4)(3)} \\
&= 2\sqrt{4}\sqrt{3} \\
&= 2(2)\sqrt{3} \\
&= 4\sqrt{3}.
\end{aligned}$$

Therefore,

$$\sqrt{48} = 4\sqrt{3}.$$

If we observe that 16 is a factor of 48, we can write directly

$$\begin{aligned}
\sqrt{48} &= \sqrt{(16)(3)} \\
&= \sqrt{16}\sqrt{3} \\
&= 4\sqrt{3}.
\end{aligned}$$

c. There are no factors of 69 that are perfect squares. Therefore, $\sqrt{69}$ cannot be simplified using Rule 1.

In reducing square roots, as in reducing fractions, you must be sure always to use *factors*. It is a common error to use addends to reduce a square root. For example, you may write $\sqrt{13} = \sqrt{4 + 9}$. But, if you separate the addends, you get $\sqrt{4} + \sqrt{9} = 2 + 3 = 5$, which is not $\sqrt{13}$. Using factors, however, it is true that $\sqrt{36} = \sqrt{(4)(9)} = \sqrt{4}\sqrt{9} = (2)(3) = 6$.

We may use Rule 2 to simplify square roots of rational numbers.

EXAMPLE 21.4   Simplify:

a. $\sqrt{\dfrac{4}{25}}$ b. $\sqrt{\dfrac{5}{16}}$ c. $\sqrt{\dfrac{12}{25}}$

Solutions   a. Using Rule 2,

$$\sqrt{\frac{4}{25}} = \frac{\sqrt{4}}{\sqrt{25}}$$

$$= \frac{2}{5}.$$

b. Five is not a perfect square. However, since 16 is a perfect square, we may write

$$\sqrt{\frac{5}{16}} = \frac{\sqrt{5}}{\sqrt{16}}$$

$$= \frac{\sqrt{5}}{4}.$$

We may also write this result in the form $\frac{1}{4}\sqrt{5}$.

c. Using Rule 2,

$$\sqrt{\frac{12}{25}} = \frac{\sqrt{12}}{\sqrt{25}}$$

$$= \frac{\sqrt{12}}{5}.$$

Then, using Rule 1,

$$\frac{\sqrt{12}}{5} = \frac{\sqrt{(4)(3)}}{5}$$

$$= \frac{\sqrt{4}\sqrt{3}}{5}$$

$$= \frac{2\sqrt{3}}{5}.$$

We may also write this result in the form $\frac{2}{5}\sqrt{3}$.

Exercise 21.1

Find the value:

1. $\sqrt{169}$ 2. $\sqrt{225}$ 3. $\sqrt{576}$ 4. $\sqrt{1089}$

5. $-\sqrt{64}$ 6. $-\sqrt{324}$ 7. $\sqrt{-144}$ 8. $\sqrt{-289}$

Simplify:

9. $\sqrt{32}$ 10. $\sqrt{72}$ 11. $\sqrt{98}$ 12. $\sqrt{500}$

13. $\sqrt{60}$ 14. $\sqrt{117}$ 15. $\sqrt{108}$ 16. $\sqrt{243}$

17. $\sqrt{\dfrac{1}{4}}$ 18. $\sqrt{\dfrac{1}{121}}$ 19. $\sqrt{\dfrac{25}{81}}$ 20. $\sqrt{\dfrac{64}{49}}$

21. $\sqrt{\dfrac{5}{36}}$ 22. $\sqrt{\dfrac{11}{9}}$ 23. $\sqrt{\dfrac{18}{121}}$ 24. $\sqrt{\dfrac{216}{25}}$

21.2   Equations of the Form $ax^2 + c = 0$

In Sections 18.1 and 18.2, we solved quadratic equations by collecting the nonzero terms on one side of the equation and factoring. However, one type of quadratic equation is easier to solve using square roots. This is the type of quadratic equation in which there is no x term.

EXAMPLE 21.5   Solve $x^2 - 4 = 0$.

Solution   First we solve for the x^2 term, writing

$$x^2 = 4.$$

Now we will take the square root of each side of the equation. Since the square of either a positive or a negative number is positive, we must allow for both the positive and negative square roots of 4. We write

$$x^2 = 4$$
$$x = \pm\sqrt{4}$$
$$x = \pm 2.$$

The solutions are 2 and -2.

Of course, we could have solved the equation in Example 21.5 by factoring:

$$x^2 - 4 = 0$$
$$(x + 2)(x - 2) = 0$$
$$x + 2 = 0 \text{ and } x - 2 = 0$$
$$x = -2 \text{ and } x = 2.$$

However, the square root method is usually faster and easier. Also, the square root method works for equations with no x term that cannot be solved by factoring over the integers.

EXAMPLE 21.6   Solve $x^2 - 7 = 0$.

Solution   First we solve for the x^2 term:

$$x^2 = 7.$$

Now we take the square root of each side of the equation:

$$x^2 = 7$$
$$x = \pm\sqrt{7}.$$

The solutions are $\sqrt{7}$ and $-\sqrt{7}$. When the solutions contain square roots, the equation could not have been solved by factoring over the integers.

We will generally simplify square roots whenever possible.

EXAMPLE 21.7   Solve $4x^2 - 3 = 0$.

Solution   Solving for the x^2 term,

$$4x^2 - 3 = 0$$
$$4x^2 = 3$$
$$x^2 = \frac{3}{4}.$$

Now we take the square root of each side, remembering that the square root on the right-hand side can be either positive or negative:

$$x^2 = \frac{3}{4}$$
$$x = \pm\sqrt{\frac{3}{4}}$$
$$= \pm\frac{\sqrt{3}}{2}.$$

The solutions are $\dfrac{\sqrt{3}}{2}$ and $-\dfrac{\sqrt{3}}{2}$.

EXAMPLE 21.8   Solve $2x^2 - 40 = 0$.

Solution

$$2x^2 - 40 = 0$$
$$2x^2 = 40$$
$$x^2 = 20$$
$$x = \pm\sqrt{20}$$
$$= \pm\sqrt{(4)(5)}$$
$$= \pm 2\sqrt{5}.$$

The solutions are $2\sqrt{5}$ and $-2\sqrt{5}$.

It is possible for an equation of the form $ax^2 + c = 0$ to have no real number solutions. This is the case if, when we take the square root of each side, the result is a square root of a negative number.

EXAMPLE 21.9   Solve $3x^2 + 9 = 0$.

Solution

$$3x^2 + 9 = 0$$
$$3x^2 = -9$$

$$x^2 = -3$$
$$x = \pm\sqrt{-3}.$$

Since we have the square root of a negative, there are no real number solutions.

☐ Exercise 21.2

Solve:

1. $x^2 - 16 = 0$

2. $x^2 - 49 = 0$

3. $x^2 - 11 = 0$

4. $x^2 - 39 = 0$

5. $16x^2 - 3 = 0$

6. $9x^2 - 10 = 0$

7. $x^2 - 75 = 0$

8. $x^2 - 56 = 0$

9. $4x^2 - 9 = 0$

10. $64x^2 - 25 = 0$

11. $9x^2 + 4 = 0$

12. $4x^2 + 36 = 0$

13. $2x^2 = 50$

14. $4x^2 = 576$

15. $25x^2 - 12 = 0$

16. $9x^2 - 32 = 0$

17. $2x^2 - 48 = 0$

18. $27x^2 - 216 = 0$

19. $4x^2 + 64 = 0$

20. $25x^2 + 625 = 0$

☐ 21.3 ☐ The Quadratic Formula

To solve quadratic equations by factoring in Sections 18.1 and 18.2, we wrote the equations in the form

$$ax^2 + bx + c = 0,$$

where all of the nonzero terms are collected on one side. The form $ax^2 + bx + c = 0$ is called the **standard form of a quadratic equation.**

A quadratic equation in standard form can be solved using the **quadratic formula.** The proof that the quadratic formula gives the solutions of the equation is given at the end of this unit.

The Quadratic Formula: For the quadratic equation in standard form,

$$ax^2 + bx + c = 0,$$

where $a \neq 0$, the solutions are given by

$$x = \frac{-b \pm \sqrt{b^2 - 4ac}}{2a}.$$

EXAMPLE 21.10   Solve $x^2 - 4x + 3 = 0$.

Solution   Since the equation is in standard form, and the expression of nonzero terms is easily factored, one way to solve the equation is by factoring:

$$x^2 - 4x + 3 = 0$$
$$(x - 1)(x - 3) = 0$$
$$x - 1 = 0 \text{ and } x - 3 = 0$$
$$x = 1 \text{ and } x = 3.$$

However, the quadratic formula will give the same solutions. Since $x^2 - 4x + 3 = 0$ is in standard form, $a = 1$, $b = -4$, and $c = 3$. Using the quadratic formula,

$$x = \frac{-b \pm \sqrt{b^2 - 4ac}}{2a},$$

we substitute 1 for a, -4 for b, and 3 for c:

$$x = \frac{-(-4) \pm \sqrt{(-4)^2 - 4(1)(3)}}{2(1)}$$

$$= \frac{4 \pm \sqrt{16 - 12}}{2}$$

$$= \frac{4 \pm \sqrt{4}}{2}$$

$$= \frac{4 \pm 2}{2}.$$

The equation $x = \dfrac{4 \pm 2}{2}$ means $x = \dfrac{4 + 2}{2}$ and $x = \dfrac{4 - 2}{2}$:

$$x = \frac{4 + 2}{2} = \frac{6}{2} = 3,$$

and

$$x = \frac{4 - 2}{2} = \frac{2}{2} = 1.$$

The solutions are 3 and 1, which, of course, are the same as those we found by factoring. Clearly, it is easier to use the method of factoring for this example.

We use the quadratic formula when we have an equation in standard form in which the expression of nonzero terms cannot be factored, or is not easily factored.

EXAMPLE 21.11   Solve $4x^2 - 8x - 21 = 0$.

Solution   This equation can be solved by factoring. However, you might not readily see the factors. We may use the quadratic formula with $a = 4$, $b = -8$, and $c = -21$:

$$x = \frac{-b \pm \sqrt{b^2 - 4ac}}{2a}$$

$$= \frac{-(-8) \pm \sqrt{(-8)^2 - 4(4)(-21)}}{2(4)}$$

$$= \frac{8 \pm \sqrt{64 + 336}}{8}$$

$$= \frac{8 \pm \sqrt{400}}{8}$$

$$= \frac{8 \pm 20}{8}.$$

Then

$$x = \frac{8 + 20}{8} = \frac{28}{8} = \frac{7}{2}$$

and

$$x = \frac{8 - 20}{8} = \frac{-12}{8} = -\frac{3}{2}.$$

The solutions are $\frac{7}{2}$ and $-\frac{3}{2}$. When $b^2 - 4ac$ is a perfect square, we have an equation that could have been solved by factoring, and we can find solutions which do not involve square roots.

If a quadratic equation is not given in standard form, we must write it in standard form in order to use the quadratic formula.

EXAMPLE 21.12 Solve $3x^2 + 11x = 4$.

Solution The equation is not written in standard form. We write

$$3x^2 + 11x - 4 = 0.$$

Then, $a = 3$, $b = 11$, and $c = -4$:

$$x = \frac{-b \pm \sqrt{b^2 - 4ac}}{2a}$$

$$= \frac{-11 \pm \sqrt{11^2 - 4(3)(-4)}}{2(3)}$$

$$= \frac{-11 \pm \sqrt{121 + 48}}{6}$$

$$= \frac{-11 \pm \sqrt{169}}{6}$$

$$= \frac{-11 \pm 13}{6}.$$

Then

$$x = \frac{-11 + 13}{6} = \frac{2}{6} = \frac{1}{3}$$

and

$$x = \frac{-11 - 13}{6} = \frac{-24}{6} = -4.$$

The solutions are $\frac{1}{3}$ and -4.

EXAMPLE 21.13   Solve $x^2 + 2x - 2 = 0$.

Solution   It will not take much experimentation for you to conclude that the expression of nonzero terms in this equation cannot be factored over the integers. Using the quadratic formula with $a = 1$, $b = 2$, and $c = -2$:

$$x = \frac{-b \pm \sqrt{b^2 - 4ac}}{2a}$$

$$= \frac{-2 \pm \sqrt{2^2 - 4(1)(-2)}}{2(1)}$$

$$= \frac{-2 \pm \sqrt{4 + 8}}{2}$$

$$= \frac{-2 \pm \sqrt{12}}{2}.$$

Recall from Section 21.1 that $\sqrt{12}$ can be simplified:

$$\sqrt{12} = \sqrt{(4)(3)} = 2\sqrt{3}.$$

Therefore,

$$x = \frac{-2 \pm \sqrt{12}}{2}$$

$$= \frac{-2 \pm 2\sqrt{3}}{2}.$$

Now, we recall from Section 19.1 that we can reduce this fraction. We must remember to divide out only factors, and not addends, so we factor the numerator:

$$x = \frac{-2 \pm 2\sqrt{3}}{2}$$

$$= \frac{2(-1 \pm \sqrt{3})}{2}$$

$$= -1 \pm \sqrt{3}.$$

The solutions are $-1 + \sqrt{3}$ and $-1 - \sqrt{3}$. It is common to leave such solutions summarized in the form

$$x = -1 \pm \sqrt{3}.$$

Since $b^2 - 4ac$ is not a perfect square, and a square root appears in the solutions, the equation could not have been solved by factoring.

When using the quadratic formula, as with the method of factoring, we may collect the nonzero terms of a quadratic equation on the right-hand side rather than the left-hand side. We might do this to make the x^2 term positive.

EXAMPLE 21.14   Solve $2 - x^2 = 4x$.

Solution   Collecting on the right, we have

$$0 = x^2 + 4x - 2,$$

and so $a = 1$, $b = 4$, and $c = -2$. Therefore,

$$x = \frac{-b \pm \sqrt{b^2 - 4ac}}{2a}$$

$$= \frac{-4 \pm \sqrt{4^2 - 4(1)(-2)}}{2(1)}$$

$$= \frac{-4 \pm \sqrt{16 + 8}}{2}$$

$$= \frac{-4 \pm \sqrt{24}}{2}$$

$$= \frac{-4 \pm \sqrt{(4)(6)}}{2}$$

$$= \frac{-4 \pm 2\sqrt{6}}{2}$$

$$= -2 \pm \sqrt{6}.$$

The solutions may be expressed as $-2 \pm \sqrt{6}$.

When $b^2 - 4ac$ is negative, a square root of a negative number will appear in the apparent solutions. In this case, the equation has no real number solutions.

EXAMPLE 21.15   Solve $2x^2 - 3x + 2 = 0$.

Solution   The equation is in standard form, and $a = 2$, $b = -3$, and $c = 2$:

$$x = \frac{-b \pm \sqrt{b^2 - 4ac}}{2a}$$

$$= \frac{-(-3) \pm \sqrt{(-3)^2 - 4(2)(2)}}{2(2)}$$

$$= \frac{3 \pm \sqrt{9 - 16}}{4}$$

$$= \frac{3 \pm \sqrt{-7}}{4}.$$

We have an expression involving the square root of a negative number, and so the equation has no real number solutions.

Observe that the number given by $b^2 - 4ac$ discriminates among equations that could be solved by factoring, equations with solutions involving a square root, and equations that have no real number solution. Therefore, $b^2 - 4ac$ is called the **discriminant** of the quadratic equation.

Exercise 21.3

Use the quadratic formula to solve:

1. $x^2 - 5x + 6 = 0$ 2. $x^2 - 3x - 18 = 0$ 3. $2x^2 - 5x + 2 = 0$ 4. $4x^2 + 4x - 3 = 0$

5. $9x^2 + 6x = 8$ 6. $4x^2 + 4 = 17x$ 7. $x^2 - 4x - 3 = 0$ 8. $x^2 + 2x - 5 = 0$

9. $4 = x^2 + 4x$ 10. $6x - x^2 = 2$ 11. $2x^2 = 2x + 1$ 12. $3x^2 + 4x = 9$

13. $7 = 2x^2 + 3x$ 14. $3x^2 = 5x + 3$ 15. $2x^2 - 5x + 5 = 0$ 16. $x^2 + x + 1 = 0$

17. $x(x - 1) = 3x + 2$ 18. $x(x + 2) = 4x(x + 1) - 3$

19. $2x(x - 3) = x(x + 3) - 3(x - 1)$ 20. $x(x - 4) = 2(x - 4) - 2$

21.4   An Application

In Section 18.3, we solved problems involving areas of rectangles. The equations used to solve these problems were quadratic equations. The problems were chosen so that the quadratic equations involved could be solved by factoring. However, it is quite possible to encounter problems of the same type that cannot be solved by factoring.

EXAMPLE 21.16   A rectangle has an area of 8 square feet. Its length is 4 feet more than its width. Find the width and length to two decimal places.

Solution   Recall that the area of a rectangle is equal to its width times its length. In this problem, the length is 4 feet more than the width; therefore, if the width is x, then the length is $x + 4$:

$$wl = A$$
$$x(x + 4) = 8$$
$$x^2 + 4x - 8 = 0.$$

Using the quadratic formula,

$$x = \frac{-b \pm \sqrt{b^2 - 4ac}}{2a}$$

$$= \frac{-4 \pm \sqrt{4^2 - 4(1)(-8)}}{2(1)}$$

$$= \frac{-4 \pm \sqrt{16 + 32}}{2}$$

$$= \frac{-4 \pm \sqrt{48}}{2}$$

$$= \frac{-4 \pm 4\sqrt{3}}{2}.$$

Dividing a factor of 2 from the numerator and the denominator,

$$x = -2 \pm 2\sqrt{3}.$$

The solution $-2 - 2\sqrt{3}$ is extraneous, because it is negative, and a rectangle cannot have a negative side. Therefore, the width is $-2 + 2\sqrt{3}$. Using either a calculator or a table of square roots (in the appendix at the end of this book),

$$\sqrt{3} \approx 1.732.$$

Therefore,

$$2\sqrt{3} \approx 3.464$$

and

$$-2 + 2\sqrt{3} \approx -2 + 3.464 = 1.464.$$

Thus, the width of the rectangle is 1.46 feet to two decimal places. (You should always carry at least one extra decimal place and round off at the end.) The length of the rectangle is 4 feet more than the width, or

$$1.464 + 4 = 5.464,$$

or 5.46 feet to two decimal places. To check, the area is $(1.46)(5.46) \approx 7.97$, or approximately 8 square feet.

EXAMPLE 21.17   Find the width x of this figure to two decimal places, if the total area is 5 square inches:

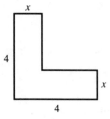

Solution   We divide the figure into two rectangles as shown by the dotted line:

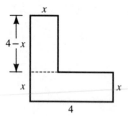

The width of the rectangle at the bottom is x, and its length is 4. Therefore, its area is

$$4x.$$

We have cut x inches off from the bottom of the figure to construct the upper rectangle. Thus, the width of the upper rectangle is x and its length is $4 - x$. Therefore, its area is

$$x(4 - x).$$

The total area of the figure is the sum of the areas of the two rectangles:

$$4x + x(4 - x) = 5.$$

Simplifying and collecting like terms,

$$4x + x(4 - x) = 5$$
$$4x + 4x - x^2 = 5$$
$$8x - x^2 = 5$$
$$0 = x^2 - 8x + 5.$$

Then, using the quadratic formula,

$$x = \frac{-b \pm \sqrt{b^2 - 4ac}}{2a}$$

$$= \frac{-(-8) \pm \sqrt{(-8)^2 - 4(1)(5)}}{2(1)}$$

$$= \frac{8 \pm \sqrt{64 - 20}}{2}$$

$$= \frac{8 \pm \sqrt{44}}{2}$$

$$= \frac{8 \pm 2\sqrt{11}}{2}$$

$$= 4 \pm \sqrt{11}.$$

Approximating $4 + \sqrt{11}$, we have

$$4 + \sqrt{11} \approx 4 + 3.317 = 7.317.$$

This solution is extraneous because it would make $4 - x$ negative. The solution is

$$4 - \sqrt{11} \approx 4 - 3.317 = 0.683,$$

or 0.68 inch to two decimal places. To check, the area of the bottom rectangle is approximately

$$4(0.68) = 2.72,$$

and the area of the upper rectangle is approximately

$$(0.68)(4 - 0.68) = (0.68)(3.32) \approx 2.26.$$

The total area is

$$2.72 + 2.26 = 4.98,$$

or approximately 5 square inches.

A figure related to the rectangle is the rectangular box. The volume of a box is given by the width times the length of its rectangular base, times the height.

EXAMPLE 21.18   The volume of a box is $2\frac{1}{2}$ cubic feet. Its length is $\frac{1}{2}$ foot more than its width, and its height is 2 feet. What are the width and length of the base of the box, to one decimal place?

Solution   A diagram of the box is

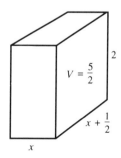

Since

$$wlh = V,$$

we have the equation

$$x\left(x + \frac{1}{2}\right)(2) = \frac{5}{2}$$

$$2x\left(x + \frac{1}{2}\right) = \frac{5}{2}$$

$$2x^2 + x = \frac{5}{2}$$

$$4x^2 + 2x = 5$$

$$4x^2 + 2x - 5 = 0.$$

Using the quadratic formula,

$$x = \frac{-b \pm \sqrt{b^2 - 4ac}}{2a}$$

$$= \frac{-2 \pm \sqrt{2^2 - 4(4)(-5)}}{2(4)}$$

$$= \frac{-2 \pm \sqrt{4 + 80}}{8}$$

$$= \frac{-2 \pm \sqrt{84}}{8}$$

$$= \frac{-2 \pm 2\sqrt{21}}{8}$$

$$= \frac{-1 \pm \sqrt{21}}{4}.$$

We rule out the negative solution as extraneous, and approximate the positive solution:

$$\frac{-1 + \sqrt{21}}{4} \approx \frac{-1 + 4.583}{4}$$

$$= \frac{3.583}{4}$$

$$= 0.896,$$

or 0.9 to one decimal place. The width is 0.9 foot, and the length is $0.9 + \frac{1}{2} = 0.9 + 0.5 = 1.4$ feet. To check, the volume is

$$(0.9)(1.4)(2) = 2.52,$$

or approximately $2\frac{1}{2}$ cubic feet.

☐_____☐ Exercise 21.4

1. The area of a rectangle is 14 square feet. The length is 2 feet more than the width. Find the width and length to two decimal places.

2. A rectangle has a length 5 inches more than its width. Its area is 5 square inches. Find its width and length to two decimal places.

3. The area of a rectangular mirror is 2 square meters. The length is $\frac{1}{2}$ meter more than the width. What are the width and length to one decimal place?

4. A rectangular rug has an area of $4\frac{1}{2}$ square meters. The width is $\frac{1}{2}$ meter less than the length. What are the width and length to one decimal place?

5. Find the width x of this figure to two decimal places, if the area is 15 square centimeters:

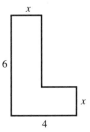

6. If the area of this figure is 30 square feet, find the width x to two decimal places:

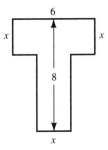

7. The base of a box is a rectangle 2 feet longer than it is wide. The height is 3 feet. The volume of the box is 6 cubic feet. What are the width and length of the base of the box to one decimal place?

8. You want to build a crate that will hold $8\frac{1}{2}$ cubic feet. Its length is to be 3 feet more than its width and its height must be $\frac{1}{2}$ foot. What should the width and length be to one decimal place?

☐ 21.5 ☐ The Parabola

Consider the quadratic equation in two variables

$$y = x^2.$$

Observe that if $x = 0$, then $y = 0$, and if $y = 0$, then $x = 0$. Thus the only y-intercept and the only x-intercept are both at $(0, 0)$. We will draw the graph of $y = x^2$ by finding some other points:

If $x = 1$, $y = 1^2 = 1$ gives $(1, 1)$.
If $x = 2$, $y = 2^2 = 4$ gives $(2, 4)$.
If $x = -1$, $y = (-1)^2 = 1$ gives $(-1, 1)$.
If $x = -2$, $y = (-2)^2 = 4$ gives $(-2, 4)$.

The graph is

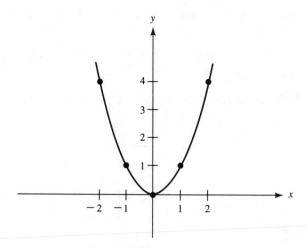

The graph is called a **parabola.** The parabola above decreases to a minimum piont, where it turns and then increases. The minimum point is called the **vertex** of the parabola. The vertex of the parabola above is clearly the point $(0, 0)$.

The vertex of a parabola also can be a maximum point. The graph of $y = -x^2$ is

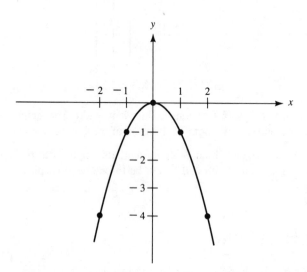

Here the vertex is again at the point $(0, 0)$, but it is a maximum; that is, the parabola increases to the vertex, where it turns and then decreases.

Consider the properties of these three parabolas:

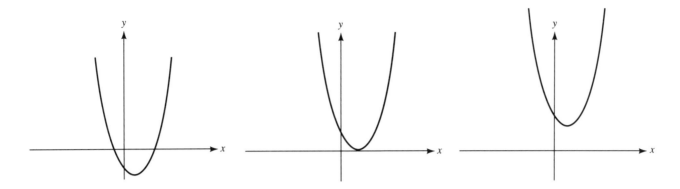

Each of the parabolas above has a vertex that is a minimum. Each has one y-intercept. However, the first has two x-intercepts, the second one, and the third none. Similarly, if the vertex is a maximum, the parabola has one y-intercept, but may have two x-intercepts, one, or none:

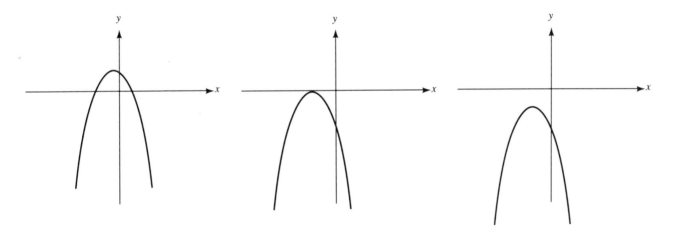

The graph of any equation of the form $y = ax^2 + bx + c$ has one y-intercept. However, the graph may have two, one, or no x-intercepts.

If the graph has two x-intercepts, and the y-intercept and x-intercepts are all distinct points, we can use these points to draw the graph.

EXAMPLE 21.19   Find the y-intercept and the x-intercepts of $y = x^2 - 9$, and draw its graph.

Solution   If $x = 0$,

$$y = 0^2 - 9 = -9.$$

Therefore, the y-intercept is $(0, -9)$. If $y = 0$,

$$0 = x^2 - 9.$$

We solve for x, using the method of Section 21.2:

$$9 = x^2$$
$$\pm \sqrt{9} = x$$
$$\pm 3 = x.$$

The x-intercepts are $(3, 0)$ and $(-3, 0)$. You should find some other points to determine that the graph is

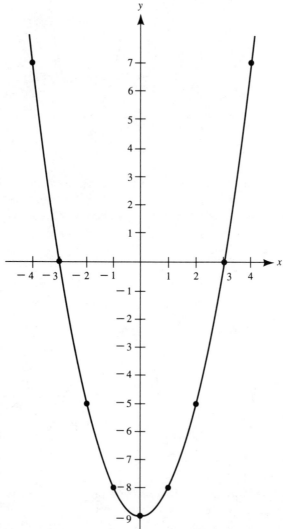

It appears that the vertex of this parabola is the point $(0, -9)$. The vertex of a parabola is an important point of the graph. In general, for a parabola given by an equation of the form

$$y = ax^2 + bx + c,$$

the x-coordinate of the vertex is given by the formula

$$x = \frac{-b}{2a}.$$

For the parabola in Example 21.19, given by the equation $y = x^2 - 9$, $a = 1$ and $b = 0$:

$$x = \frac{-b}{2a}$$

$$x = \frac{-0}{2(1)}$$

$$x = 0.$$

When $x = 0$, we have already found that $y = -9$. Therefore, the vertex of the parabola is $(0, -9)$.

EXAMPLE 21.20   Find the y-intercept, x-intercepts, and vertex for $y = x^2 - 4x + 3$, and draw its graph.

Solution   If $x = 0$,

$$y = 0^2 - 4(0) + 3 = 3,$$

so the y-intercept is $(0, 3)$. If $y = 0$,

$$0 = x^2 - 4x + 3$$
$$0 = (x - 1)(x - 3)$$
$$x - 1 = 0 \text{ and } x - 3 = 0$$
$$x = 1 \text{ and } x = 3.$$

The x-intercepts are $(1, 0)$ and $(3, 0)$. To find the vertex, we use

$$x = \frac{-b}{2a}$$
$$= \frac{-(-4)}{2(1)}$$
$$= \frac{4}{2}$$
$$= 2.$$

Then, when $x = 2$,

$$y = x^2 - 4x + 3$$
$$= 2^2 - 4(2) + 3$$
$$= 4 - 8 + 3$$
$$= -1.$$

The vertex is $(2, -1)$. The graph is

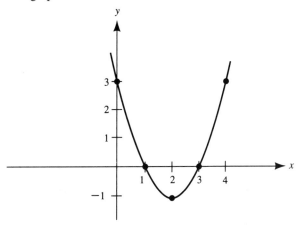

Observe that the points on each side of the vertex are symmetrical; that is, the points to the right of the vertex are mirror images of the points to the left of the vertex. For example, the point $(4, 3)$ appears to be the mirror image of the point $(0, 3)$. If $x = 4$,

$$
\begin{aligned}
y &= x^2 - 4x + 3 \\
&= 4^2 - 4(4) + 3 \\
&= 16 - 16 + 3 \\
&= 3.
\end{aligned}
$$

Thus, the point $(4, 3)$ is on the parabola as shown. You may use such a point as a check point.

The x-intercepts of a parabola may involve square roots. We can find them using the quadratic formula.

EXAMPLE 21.21   Find the y-intercept, x-intercepts, and vertex for $y = x^2 + 2x - 4$, and draw the graph.

Solution   The y-intercept is $(0, -4)$. To find the x-intercepts, if $y = 0$ we have

$$
0 = x^2 + 2x - 4.
$$

Using the quadratic formula,

$$
\begin{aligned}
x &= \frac{-2 \pm \sqrt{4 - 4(1)(-4)}}{2(1)} \\
&= \frac{-2 \pm \sqrt{4 + 16}}{2} \\
&= \frac{-2 \pm \sqrt{20}}{2} \\
&= \frac{-2 \pm 2\sqrt{5}}{2} \\
&= -1 \pm \sqrt{5}.
\end{aligned}
$$

We may use a calculator or table of square roots to approximate these solutions:

$$
\begin{aligned}
-1 + \sqrt{5} &\approx -1 + 2.2 = 1.2 \\
-1 - \sqrt{5} &\approx -1 - 2.2 = -3.2,
\end{aligned}
$$

so the x-intercepts are approximately $(1.2, 0)$ and $(-3.2, 0)$. The x-coordinate of the vertex is

$$
\begin{aligned}
x &= \frac{-b}{2a} \\
&= \frac{-2}{2(1)} \\
&= -1.
\end{aligned}
$$

If $x = -1$,

$$y = x^2 + 2x - 4$$
$$= (-1)^2 + 2(-1) - 4$$
$$= 1 - 2 - 4$$
$$= -5.$$

Therefore, the vertex is $(-1, -5)$. The graph is

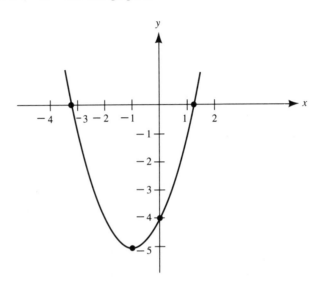

If, in the quadratic formula, the number under the square root is negative, then the parabola has no x-intercepts.

EXAMPLE 21.22   Find the y-intercept, x-intercepts, and vertex for $y = x^2 - 2x + 3$, and draw the graph.

Solution   The y-intercept is $(0, 3)$. If $y = 0$,

$$0 = x^2 - 2x + 3,$$

and

$$x = \frac{2 \pm \sqrt{4 - 4(1)(3)}}{2(2)}$$

$$= \frac{2 \pm \sqrt{-8}}{4}.$$

Therefore, there are no x-intercepts. The x-coordinate of the vertex is

$$x = \frac{-b}{2a}$$

$$= \frac{-(-2)}{2(1)}$$

$$= \frac{2}{2}$$

$$= 1.$$

If $x = 1$,

$$y = x^2 - 2x + 3$$
$$= 1^2 - 2(1) + 3$$
$$= 1 - 2 + 3$$
$$= 2.$$

The vertex is $(1, 2)$.

We have found only two points on the parabola, the y-intercept and the vertex. To construct the graph, we choose values of x on each side of the vertex to find other points:

If $x = 2$, $y = 2^2 - 2(2) + 3 = 3$ gives $(2, 3)$.
If $x = 3$, $y = 3^2 - 2(3) + 3 = 6$ gives $(3, 6)$.
If $x = -1$, $y = (-1)^2 - 2(-1) + 3 = 6$ gives $(-1, 6)$.

The graph is

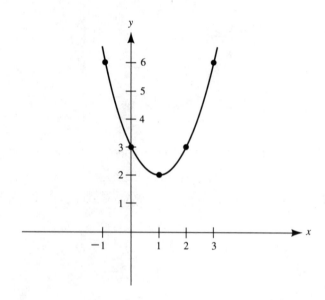

Finally, a parabola may have just one x-intercept. In this case, the x-intercept comes from a double root of a quadratic equation. Moreover, the vertex will be the same point as the one x-intercept.

EXAMPLE 21.23   Find the y-intercept, x-intercepts, and vertex for $y = x^2 + 4x + 4$, and draw the graph.

Solution   The y-intercept is $(0, 4)$. If $y = 0$,

$$0 = x^2 + 4x + 4$$
$$0 = (x + 2)(x + 2)$$
$$x + 2 = 0$$
$$x = -2.$$

There is one x-intercept, $(-2, 0)$. The x-coordinate of the vertex is

$$x = \frac{-b}{2a}$$

$$= \frac{-4}{2(1)}$$
$$= -2.$$

Therefore, the vertex is the same point as the x-intercept.

We use the y-intercept and the point that is both the vertex and x-intercept, and find a few other points on each side of the vertex to draw the parabola. The graph is

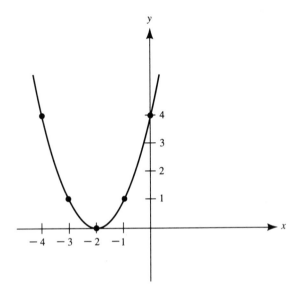

If a is negative in an equation of the form $y = ax^2 + bx + c$, the vertex of the corresponding parabola is a maximum. We have the same three cases: there can be two, one, or no x-intercepts. We use the same methods to draw the graph.

EXAMPLE 21.24   Find the y-intercept, x-intercepts, and vertex for $y = -x^2 - 2x + 8$, and draw the graph.

Solution   The y-intercept is $(0, 8)$. If $y = 0$,

$$0 = -x^2 - 2x + 8$$
$$x^2 + 2x - 8 = 0$$
$$(x + 4)(x - 2) = 0$$
$$x + 4 = 0 \text{ and } x - 2 = 0$$
$$x = -4 \text{ and } x = 2.$$

The x-intercepts are $(-4, 0)$ and $(2, 0)$. The x-coordinate of the vertex is

$$x = \frac{-b}{2a}$$
$$= \frac{-(-2)}{2(-1)}$$
$$= \frac{2}{-2}$$
$$= -1.$$

If $x = -1$,

$$y = -x^2 - 2x + 8$$
$$= -(-1)^2 - 2(-1) + 8$$
$$= -1 + 2 + 8$$
$$= 9.$$

The vertex is $(-1, 9)$. The graph is

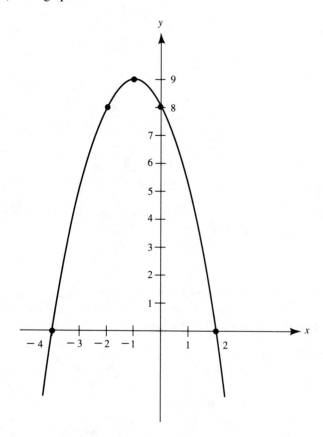

Exercise 21.5

Find the y-intercept, x-intercepts, and vertex, and draw the graph:

1. $y = x^2 - 4$

2. $y = x^2 - 1$

3. $y = x^2 - 6x + 8$

4. $y = x^2 - 2x - 3$

5. $y = x^2 - 2x - 2$

6. $y = x^2 - 4x - 3$

7. $y = x^2 - 3$

8. $y = 2x^2 - 10$

9. $y = x^2 - 2x + 2$

10. $y = x^2 + x + 1$

11. $y = x^2 + 2x + 1$

12. $y = x^2 - 6x + 9$

13. $y = -x^2 + 2x + 15$

14. $y = -x^2 - 4x + 5$

15. $y = 3 - 2x - 2x^2$

16. $y = 2x - 2x^2 - 1$

Proofs of Rules for Square Roots

From the definition of a square root, if we let

$$\sqrt{a} = x \text{ and } \sqrt{b} = y,$$

then

$$a = x^2 \text{ and } b = y^2.$$

Rule 1. Using the definitions above,

$$\sqrt{a}\sqrt{b} = xy \text{ and } ab = x^2y^2.$$

But also, using rules for exponents,

$$ab = x^2y^2 = (xy)^2,$$

and so

$$\sqrt{ab} = xy \text{ because } ab = (xy)^2.$$

Therefore,

$$\sqrt{ab} = \sqrt{a}\sqrt{b}.$$

Rule 2. Using the definitions above,

$$\frac{\sqrt{a}}{\sqrt{b}} = \frac{x}{y} \text{ and } \frac{a}{b} = \frac{x^2}{y^2}.$$

But also, using rules for exponents,

$$\frac{a}{b} = \frac{x^2}{y^2} = \left(\frac{x}{y}\right)^2,$$

and so

$$\sqrt{\frac{a}{b}} = \frac{x}{y} \text{ because } \frac{a}{b} = \left(\frac{x}{y}\right)^2.$$

Therefore,

$$\sqrt{\frac{a}{b}} = \frac{\sqrt{a}}{\sqrt{b}}.$$

Proof of the Quadratic Formula

To prove the quadratic formula, we need a method called **completing the square.** Consider the expressions

$$x^2 + 2x + 1$$
$$x^2 + 4x + 4$$
$$x^2 + 6x + 9$$
$$x^2 - 2x + 1$$
$$x^2 - 4x + 4$$

and so on. These trinomials are perfect squares because

$$x^2 + 2x + 1 = (x + 1)^2$$
$$x^2 + 4x + 4 = (x + 2)^2$$
$$x^2 + 6x + 9 = (x + 3)^2$$
$$x^2 - 2x + 1 = (x - 1)^2$$
$$x^2 - 4x + 4 = (x - 2)^2$$

and so on.

Completing the square means, given an x^2 term with coefficient 1 and an x term, to find a constant term that will make a trinomial that is a perfect square. For example, if we have

$$x^2 - 6x,$$

the completed square is

$$x^2 - 6x + 9$$

since

$$x^2 - 6x + 9 = (x - 3)^2.$$

To find the constant term, we take one-half of the coefficient of the x term, and then square the result. Given

$$x^2 - 6x,$$

we take $\frac{1}{2}(-6) = -3$, and $(-3)^2 = 9$. Given

$$x^2 + 10x,$$

we take $\frac{1}{2}(10) = 5$, and $5^2 = 25$. In this case the completed square is

$$x^2 + 10x + 25 = (x + 5)^2.$$

Observe that the number in the right-hand expression is one-half of the coefficient of the x term.

This method also works for coefficients that are odd numbers or fractions. To complete the square for

$$x^2 + 7x,$$

we take $\frac{1}{2}(7) = \frac{7}{2}$, and $\left(\frac{7}{2}\right)^2 = \frac{49}{4}$. The completed square is

$$x^2 + 7x + \frac{49}{4} = \left(x + \frac{7}{2}\right)^2.$$

To complete the square for

$$x^2 - \frac{3}{2}x,$$

we take $\frac{1}{2}\left(-\frac{3}{2}\right) = -\frac{3}{4}$, and $\left(-\frac{3}{4}\right)^2 = \frac{9}{16}$. The completed square is

$$x^2 - \frac{3}{2}x + \frac{9}{16} = \left(x - \frac{3}{4}\right)^2.$$

Now, to prove the quadratic formula, we start with the quadratic equation in standard form.

$$ax^2 + bx + c = 0.$$

We make the coefficient of the x^2 term 1 by dividing each term by a:

$$x^2 + \frac{b}{a}x + \frac{c}{a} = 0.$$

The completion of the square for $x^2 + \frac{b}{a}x$ is $\frac{1}{2}\left(\frac{b}{a}\right) = \frac{b}{2a}$, and $\left(\frac{b}{2a}\right)^2 = \frac{b^2}{4a^2}$. Adding $\frac{b^2}{4a^2}$ to each side, we have

$$x^2 + \frac{b}{a}x + \frac{b^2}{4a^2} + \frac{c}{a} = \frac{b^2}{4a^2},$$

and substituting the perfect square,

$$\left(x + \frac{b}{2a}\right)^2 + \frac{c}{a} = \frac{b^2}{4a^2}.$$

Subtracting $\frac{c}{a}$ from both sides,

$$\left(x + \frac{b}{2a}\right)^2 = \frac{b^2}{4a^2} - \frac{c}{a}$$

$$\left(x + \frac{b}{2a}\right)^2 = \frac{b^2}{4a^2} - \frac{4ac}{4a^2}$$

$$\left(x + \frac{b}{2a}\right)^2 = \frac{b^2 - 4ac}{4a^2}.$$

Now, we take the square root of each side:

$$x + \frac{b}{2a} = \pm\sqrt{\frac{b^2 - 4ac}{4a^2}}.$$

Finally, solving for x and simplifying,

$$x = -\frac{b}{2a} \pm \frac{\sqrt{b^2 - 4ac}}{2a}$$

or

$$x = \frac{-b \pm \sqrt{b^2 - 4ac}}{2a},$$

which is the quadratic formula.

Self-test

1. Reduce:

   a. $\sqrt{200}$

   b. $\sqrt{\dfrac{3}{64}}$

   1a. _____

   1b. _____

Solve for x:

2. $2x^2 - 16 = 0$

   2. _____

3. $4x = 2x^2 - 7$

   3. _____

4. A rectangle has an area of 12 square inches. Its length is 3 inches more than its width. Find its width and length to one decimal place.

   4. _____

5. Find the y-intercept, x-intercepts, and vertex for $y = x^2 + 4x + 3$, and draw the graph.

   y-intercept _____

   x-intercepts _____

   vertex _____

Answers

Unit 1

Exercise 1.1

Constants are listed on the first line. Variables are listed on the second line:

1. 3
   x

2. 25
   x

3. None
   u, v

4. None
   y, z

5. $\dfrac{5}{3}$
   t, u

6. $\dfrac{1}{2}$
   x, y, z

7. $\dfrac{3}{4}, \dfrac{2}{5}$
   u, v

8. 10.2, 5.4
   x, y

9. $\dfrac{1}{2}$
   x, y

10. $\dfrac{2}{3}$
    p, q

11. 9.3
    x, y, z

12. 10.2
    r, s

13. 2, 3
    s, t

14. None
    s, t, u, v

15. 2

16. 2

17. 2

18. 2

19. 3

20. 4

21. 2

22. 2

23. 1

24. 1

25. 1

26. 1

27. 2

28. 2

29. 3

30. 4.4

31. $\dfrac{1}{3}$

32. 6.4

33. $\dfrac{1}{4}$

34. 5.6

35. $\dfrac{1}{3}$

36. $\dfrac{2}{3}$

37. $\dfrac{1}{2}$, 1.3

38. 12.4, $\dfrac{2}{3}$

39. $\dfrac{1}{4}$, 1

40. 1, 1

Exercise 1.2

1. 75

2. 54

3. $\dfrac{73}{9}$

4. $\dfrac{10}{3}$

5. $\dfrac{3}{4}$

6. $\dfrac{7}{24}$

7. 14

8. 26

9. $\dfrac{14}{15}$

10. $\dfrac{41}{120}$

11. 47

12. 24.6

13. 95.3

14. 3.159

15. 5.67

16. 11.35

Exercise 1.3

1. Commutative property

2. Associative property

3. Associative property

4. Commutative property

5. Commutative property

6. Commutative property

7. Associative property

8. Commutative property

9. Commutative property

10. Associative property

11. Both

12. Both

Exercise 1.4

1. $3p + 3q$

2. $\dfrac{1}{2}a + \dfrac{1}{2}b$

3. $ax + ay$

4. $2sx + 2tx$

5. $15a + 10b$

6. $96y + 112z$

7. $\dfrac{1}{2}m + 2n$

8. $4u + \dfrac{15}{4}v$

9. $3a + \dfrac{5}{2}b + 6c + \dfrac{5}{4}d$

10. $\dfrac{9}{25}w + \dfrac{3}{10}x + \dfrac{1}{6}y + \dfrac{1}{5}z$

11. $4p + \dfrac{9}{2}q + 5r$

12. $\dfrac{16}{3}t + \dfrac{8}{3}u + \dfrac{24}{5}v + \dfrac{8}{5}w$

13. $5x + 10y$

14. $16r + 28s$

15. $3.96x + 2.76y + 9.46z + 2.86u$

16. $17.64p + 26.264q + 36.456r$

Self-test

1. Constants: $3, \dfrac{1}{2}, 2.5$

   Variables: x, y (Objective 1)

2. Coefficient of x: $\dfrac{2}{3}$

   Coefficient of y: 3.3 (Objective 1)

3. $\dfrac{47}{18}$ (Objective 2)

4. $x + \dfrac{5}{2}y + 2u + \dfrac{4}{3}v$ (Objective 4)

5a. Associative property

5b. Commutative property (Objective 3)

Unit 2

Exercise 2.1

1. Solution
2. Not a solution
3. Not a solution
4. Solution
5. Solution
6. Not a solution
7. Solution
8. Solution
9. Solution
10. Not a solution
11. Solution
12. Solution
13. Not a solution
14. Solution
15. Not a solution
16. Solution

Exercise 2.2

1. 20
2. 12
3. 3.2
4. 0.8
5. $\frac{1}{4}$
6. $\frac{3}{10}$
7. 22
8. 25
9. 6.1
10. 35
11. $\frac{8}{15}$
12. $\frac{31}{20}$
13. $\frac{9}{16}$
14. $\frac{18}{25}$
15. $\frac{43}{24}$
16. $\frac{26}{45}$
17. 11.22
18. 0.9
19. 0
20. 5

Exercise 2.3

1. 4
2. 16
3. $\frac{3}{5}$
4. $\frac{3}{2}$
5. 4
6. 0.9
7. $\frac{3}{2}$
8. 4
9. 45
10. 99
11. 3.9
12. 4.48
13. 20
14. $\frac{6}{5}$
15. $\frac{10}{3}$
16. 8
17. $\frac{25}{2}$
18. $\frac{75}{64}$
19. $\frac{16}{9}$
20. 1

Exercise 2.4

1. 13.77
2. 0.07625
3. 120
4. 85.6
5. 16.2%
6. 85.75%
7. 25%
8. 33.33%

9. $8.00

10. $899.99

11. $1\frac{2}{3}$ hours

12. $1\frac{1}{4}$ or 1.25 hours

13. 6 knots (nautical miles per hour)

14. 250 feet per second

15. 2.92¢

16. $0.0366 or 3.66¢

17. $1.16

18. 244¢ or $2.44

19. 55¢

20. $5.31

Self-test

1a. Not a solution

1b. Solution   (Objective 1)

2. $\frac{3}{2}$ (Objective 3)

3. 2.1   (Objective 2)

4. $\frac{2}{15}$ (Objective 2)

5. $42.50   (Objective 4)

Unit 3

Exercise 3.1

1.

2.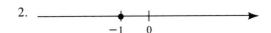

3.

4.

5.

6.

7.

8.

9.

10.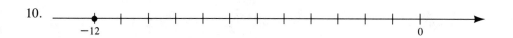
-12 0

11. 4   12. -3 13. 7   14. 10

15. -8 16. -10

Exercise 3.2

1. $>$ 2. $<$ 3. $>$ 4. $<$

5. $<$ 6. $>$ 7. $>$ 8. $<$

9. $<$ 10. $>$ 11. $>$ 12. $>$

13. $<$ 14. $<$ 15. $>$ 16. $<$

Exercise 3.3

1. 7   2. 12   3. 7   4. 12

5. 0   6. 0   7. 40   8. 100

9. 12   10. 45   11. 128   12. 1200

13. 3   14. 13   15. 48   16. 30

17. 0   18. 44   19. 0   20. 36

21. 52   22. 12   23. 8   24. 20

25. 8   26. 12   27. 7   28. 12

29. 1   30. 0

Self-test

1a.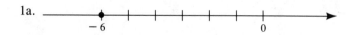
-6 0

1b.    (Objective 1)
0   4

2a. 3   2b. -4 (Objective 1)

3a. $<$ 3b. $>$ 3c. $<$ 3d. $>$ (Objective 2)

4a. 7   4b. 11 (Objective 3)   5a. 5   5b. 6 (Objective 3)

▭ Unit 4

▭ Exercise 4.1

1. -12	2. -19	3. -43	4. -104
5. 4	6. -4	7. -5	8. 9
9. -7	10. 8	11. -6	12. 11
13. 7	14. -3	15. 13	16. -33
17. -11	18. -25	19. 0	20. 0

▭ Exercise 4.2

1. -45	2. -72	3. -64	4. -77
5. 0	6. 0	7. 0	8. 0
9. 65	10. 80	11. 84	12. 225
13. -28	14. -20	15. 192	16. 91
17. -26	18. -24	19. 144	20. 135

▭ Exercise 4.3

1. -3	2. -8	3. -25	4. -1
5. 6	6. 2	7. 1	8. 100
9. 0	10. 0	11. Undefined	12. Undefined

▭ Exercise 4.4

1. 12	2. -23	3. -14	4. -18
5. -32	6. -10	7. 12	8. 21
9. 13	10. 33	11. 4	12. -9
13. -8	14. 1	15. 30	16. 0
17. 6	18. 20	19. -12	20. 16
21. 60	22. 42	23. -11	24. 19
25. 9	26. 10	27. 4	28. 6

29. 15°

30. 23°

31. 20°

32. 3°

33. 160 feet

34. 41 meters

35. 8000 feet

36. 16 fathoms

| Exercise 4.5

1. 5

2. −8

3. −10

4. 3

5. 12

6. 16

7. 20

8. −14

9. −52

10. −27

11. −2

12. 0

13. −7

14. 2

15. 0

16. Undefined

17. 1

18. 21

19. −20

20. −35

| Self-test

1a. −39   (Objective 1)

1b. −6   (Objective 1)

1c. −20   (Objective 4)

1d. 32   (Objective 4)

2a. 42   (Objective 2)

2b. −8   (Objective 3)

2c. 0   (Objective 2)

2d. Undefined   (Objective 3)

3a. 88   (Objective 2)

3b. 3   (Objective 4)

4. −8   (Objective 5)

5. 31°   (Objective 4)

| Unit 5

| Self-test

1. $\dfrac{23}{24}$ (Unit 1)

2. $\dfrac{3}{2}s + t + \dfrac{1}{2}u + 2v$ (Unit 1)

3. $\dfrac{3}{10}$ (Unit 2)

4. $\dfrac{7}{2}$ (Unit 2)

5. $7.43   (Unit 2)

6a.

6b.

(Unit 3)

7. 7   (Unit 3)

8a. -14

8b. -95 (Unit 4)

9a. 27

9b. 0   (Unit 4)

10. 56   (Unit 4)

Unit 6

Exercise 6.1

1. $11y$
2. $19p$
3. $10z$
4. $16u$

5. $3x + 4y$
6. $8y - 2z$
7. $6p$
8. $-r$

9. $-9x$
10. $10s$
11. v
12. $-4t$

13. $6x - 3y$
14. $-2x - 10$
15. $2ax - 7bx + 12$
16. $-y - by$

17. $x - 3xy + 2y$
18. $6x + 5xy - 6y$
19. $-5ab$
20. $4abc$

Exercise 6.2

1. $12x + 20$
2. $18y + 54$
3. $28z - 14$

4. $24 - 8t$
5. $27p - 30q$
6. $12r - 48s$

7. $10x - 15y - 50$
8. $44u + 66v - 121$
9. $-24s + 30$

10. $-60t + 25$
11. $-144 + 36t$
12. $-72x + 63y$

13. $-24x + 32z + 48$
14. $-10x + 20y - 30w$
15. $-4r + 9$

16. $-12 + 16y$
17. $-6x + 3$
18. $3 + 4y$

19. $-2x - 4y + 9$
20. $5x + 6y - 10$

Exercise 6.3

1. $12x - 5$
2. $9t - 18$
3. $-2r$
4. 20

5. $3y + 2z - 11$
6. $9x - y - 4z$
7. $4x + 5$
8. $-s - 21$

9. $-5y$
10. 6
11. $-6u - 15$
12. $-5s + 4t + 1$

13. $5x - 5$
14. $4y + 23$
15. $4x - 3$
16. $-r + 24$

17. $2x - 4y - 1$
18. $6x - 3$
19. $-xy + 4y$
20. $x + 2xy + y$

Exercise 6.4

1. $4y$
2. $25z + 40$
3. $-8s + 14$
4. $-26x - 12$

5. $12t - 8$
6. $8y - 64$
7. $3t - 6$
8. $7y + 8$

9. $-9x + 30$
10. $21z + 25$
11. $10r - 40$
12. $25s + 70$

13. $-28p + 30$
14. -30
15. $-36y - 35$
16. $-15x$

Self-test

1. $-4u$ (Objective 1)
2. $5x + 2xy - 2y$ (Objective 1)

3. $13t - 3$ (Objective 3)
4. $-15x + 5y - 20$ (Objective 2)

5. $-7x - 17$ (Objective 4)

Unit 7

Exercise 7.1

1. 8
2. -6
3. 2
4. -3

5. 8
6. 5
7. -4
8. -9

9. 1
10. 4
11. -3
12. -6

13. $\frac{8}{3}$
14. $\frac{9}{10}$
15. 30
16. -10

17. 3
18. $\frac{7}{2}$
19. -1
20. $-\frac{3}{2}$

21. $-\frac{3}{4}$
22. $-\frac{1}{2}$
23. $\frac{3}{2}$
24. $\frac{9}{4}$

Exercise 7.2

1. 1
2. 4
3. $\frac{11}{5}$
4. $\frac{3}{4}$

5. -8
6. $\frac{1}{3}$
7. $\frac{1}{2}$
8. -2

9. 13
10. -1
11. 0
12. 0

13. No solution
14. All real numbers
15. All real numbers
16. No solution

☐ **Exercise 7.3**

1. 15
2. -12
3. $\dfrac{15}{2}$
4. 3

5. -9
6. -16
7. $-\dfrac{28}{3}$
8. $\dfrac{20}{7}$

9. $\dfrac{7}{5}$
10. -5
11. 6
12. $-\dfrac{3}{4}$

13. $-\dfrac{1}{9}$
14. 4
15. No solution
16. All real numbers

☐ **Self-test**

1. $-\dfrac{7}{4}$ (Objective 1)
2. No solution   (Objective 3)

3. -3 (Objective 3)
4. $\dfrac{3}{2}$ (Objective 2)

5. All real numbers   (Objective 2)

☐ # Unit 8

☐ **Exercise 8.1**

1. $x + 14$
2. $x + 3$
3. $x - 5$
4. $x - 30$

5. $7 - x$
6. $x - 7$
7. $9 - x$
8. $200 - x$

9. $3x$
10. $10x$
11. $\dfrac{1}{3}x$
12. $\dfrac{5}{2}x$

13. $\dfrac{2}{3}x + 18$
14. $\dfrac{1}{5}x - 10$
15. $\dfrac{2}{5}(x + 8)$
16. $\dfrac{3}{4}(x - 12)$

17. 8
18. $\dfrac{1}{2}$
19. 6
20. -15

21. 29
22. 53
23. 23
24. -5

☐ **Exercise 8.2**

1. $84°$
2. $112°$
3. $58°, 68°, 54°$

4. $77°, 57°, 46°$
5. $49°, 33°, 98°$
6. $30°, 60°, 90°$

7. 66°, 44°, 70°

8. 40°, 60°, 80°

9. 22.5°, 67.5°, 90°

10. 36°, 48°, 96°

11. 12 feet, 20 feet, 24 feet

12. 11 meters, 8 meters, 14 meters

13. 5 yards, 15 yards, 15 yards

14. 22 inches, 18 inches, 36 inches

15. 30 feet, 20 feet

16. 23 centimeters, 45 centimeters

17. 16 inches, 8 inches

18. 2 yards, $1\frac{1}{3}$ yards

19. 10 feet, $12\frac{1}{2}$ feet

20. $2\frac{2}{3}$ meters, $5\frac{1}{3}$ meters

Exercise 8.3

1. 23 pennies, 5 dimes

2. 9 pennies, 11 nickels

3. 22 nickels, 18 dimes

4. 19 dimes, 17 quarters

5. 24 quarters, 17 half dollars

6. 4 nickels, 16 quarters

7. 45 pennies, 39 nickels, 34 dimes

8. 38 nickels, 41 dimes, 19 quarters

9. 80 for closer rows, 110 for back rows

10. 260 student tickets, 212 guest tickets

11. 22 packages potato chips, 28 packages pretzels

12. 54 cups of coffee, 30 cups of tea

13. 2 pounds Bosc, 3 pounds Bartlett

14. 25 pounds fine sand, 25 pounds coarse sand

15. 20 ball-point pens, 10 felt-tip pens

16. 15 cans light tuna, 9 cans white tuna

17. 4 pounds kidney beans, 6 pounds green beans

18. 60 ounces beef flavor, 20 ounces liver flavor

19. $7\frac{1}{2}$ pecks peat moss, $17\frac{1}{2}$ pecks potting soil

20. $22\frac{1}{2}$ quarts olive oil, $27\frac{1}{2}$ quarts vegetable oil

Exercise 8.4

1. 160

2. 90

3. 5 gallons

4. 1700 miles

5. 4 inches

6. $87\frac{1}{2}$ miles

7. 6 feet

8. $22\frac{1}{2}$ pounds

9. 40.64 centimeters

10. 39.37 inches

11. 54 miles

12. 1.855 quarts

Self-test

1. $\frac{4}{5}x - 5$ (Objective 1)

2. $22\frac{1}{2}$ meters   (Objective 4)

3. 9 dimes, 11 nickels   (Objective 3)

4. $12\frac{1}{2}$ liters ginger ale, $2\frac{1}{2}$ liters fruit concentrate   (Objective 3)

5. $2\frac{1}{2}$ feet, $4\frac{1}{2}$ feet   (Objective 2)

⬚ Unit 9

⬚ Exercise 9.1

1. 8

2. 6

3. 66

4. -58

5. 3

6. $-\dfrac{8}{3}$

7. $400

8. $187.50

9. $515

10. $873.75

11. 16.2

12. 31.5

13. 15.7

14. 29.8

15. -5

16. -25

17. 5

18. -58

19. $\dfrac{12}{7}$

20. $\dfrac{18}{5}$ or 3.6

⬚ Exercise 9.2

1. $R = \dfrac{P}{B}$

2. $l = \dfrac{A}{w}$

3. $R = \dfrac{I}{PT}$

4. $h = \dfrac{2A}{b}$

5. $E = IR$

6. $F = kx$

7. $R = \dfrac{E}{I}$

8. $w = \dfrac{F}{\mu}$

9. $t = \dfrac{v - v_0}{a}$

10. $l = \dfrac{P - 2w}{2}$

11. $y = \dfrac{9 - 2x}{5}$

12. $y = \dfrac{-6x - 4}{3}$

13. $y = \dfrac{4x - 5}{6}$

14. $y = \dfrac{3x + 7}{4}$

15. $y = \dfrac{2x + 10}{3}$

16. $y = \dfrac{8 - 5x}{6}$

17. $P = \dfrac{A}{1 + rt}$

18. $h = \dfrac{2A}{B + b}$

19. $t = \dfrac{A - P}{Pr}$

20. $B = \dfrac{2A - bh}{h}$

⬚ Exercise 9.3

1. $W = kd$

2. $c = kr$

3. $a = \dfrac{k}{m}$

4. $I = \dfrac{k}{R}$

5. $S = krh$

6. $G = kM_1M_2$

7. $R = \dfrac{kl}{A}$

8. $P = \dfrac{kF}{A}$

9. $D = 44T$

10. $F = -\dfrac{1}{2}x$

11. $R = \dfrac{60}{T}$

12. $y = \dfrac{50}{x}$

13. $z = -2xy$

14. $A = \dfrac{1}{2}bh$

15. $p = \dfrac{4m}{n}$

16. $P = \dfrac{20.8t}{V}$

17. 165

18. 12

19. $-\dfrac{15}{4}$

20. 80

21. 75 dynes

22. 10.625 centimeters

23. 70.56 joules

24. 0.04 ohms

Self-test

1. 17   (Objective 1)

2. 14°   (Objective 1)

3. $r = \dfrac{A - P}{Pt}$ (Objective 2)

4. $y = \dfrac{12 - 5x}{6}$ (Objective 2)

5. −3   (Objective 3)

Unit 10

Exercise 10.1

1.

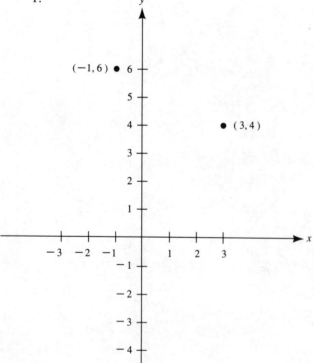

2.

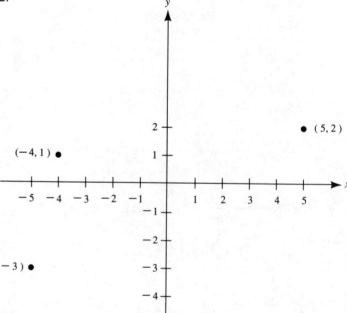

3.

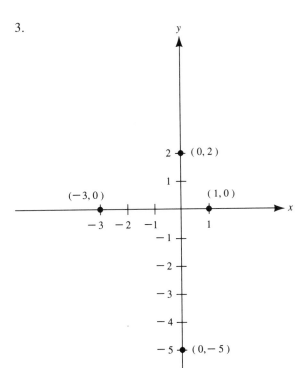

4.

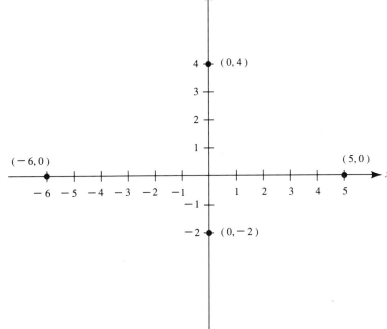

5.

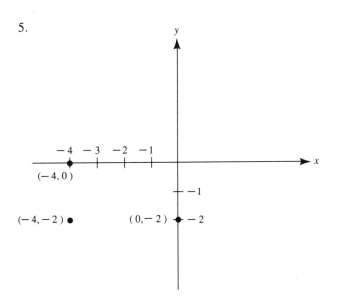

6.

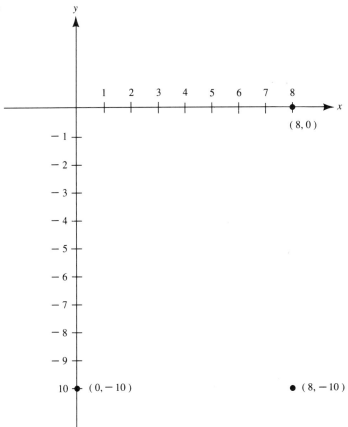

Exercise 10.2

1.

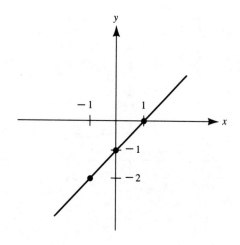

2.

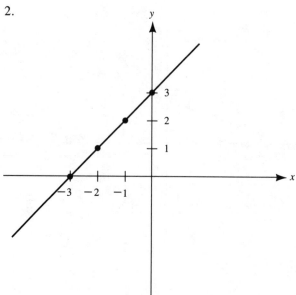

3.

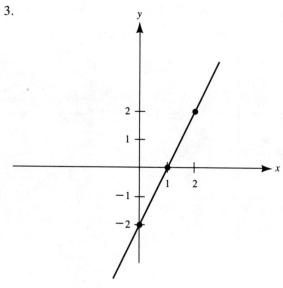

4.

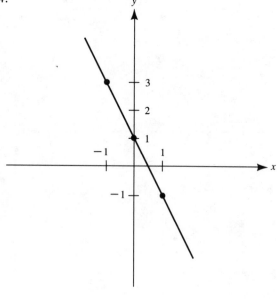

5.

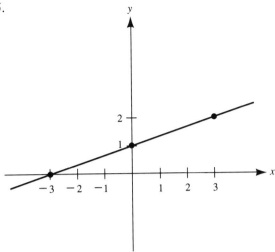

6.

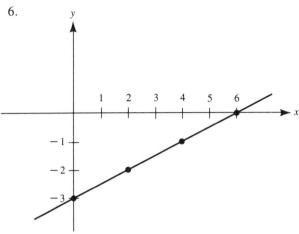

7.

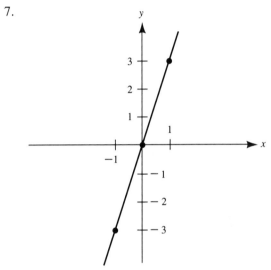

8.

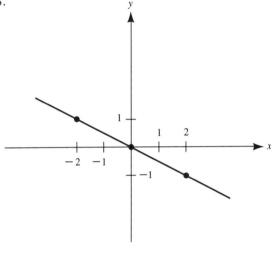

9.

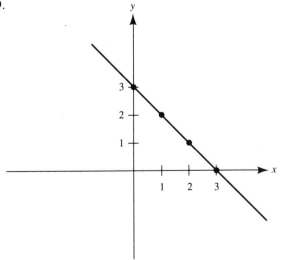

10.

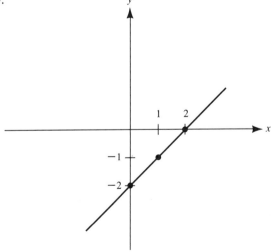

11.

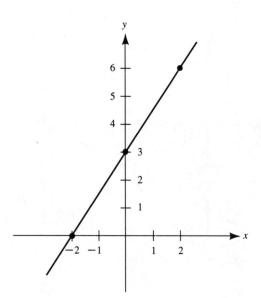

12.

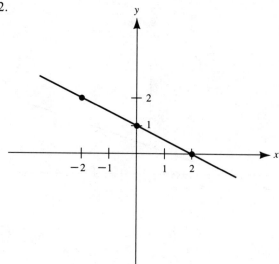

13.

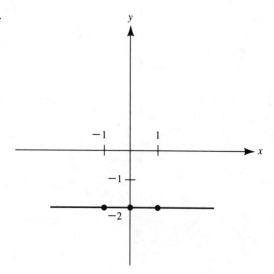

14.

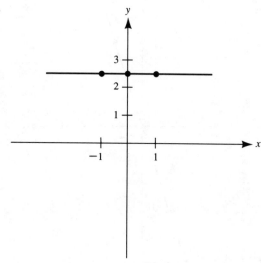

15.

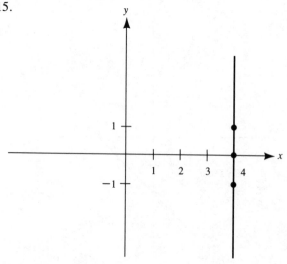

16.

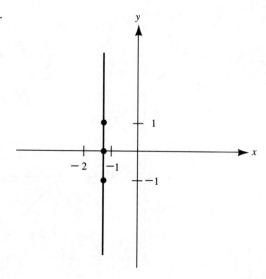

Exercise 10.3

1. (0, 5), (5, 0)

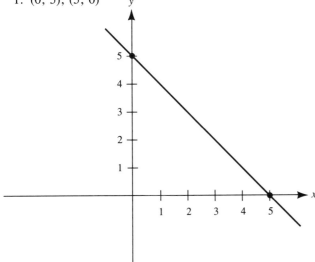

2. (0, −4), (4, 0)

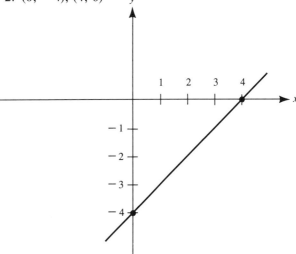

3. $\left(0, -\dfrac{3}{2}\right)$, (3, 0)

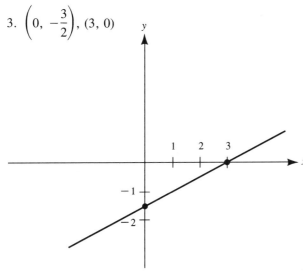

4. (0, 1), (−6, 0)

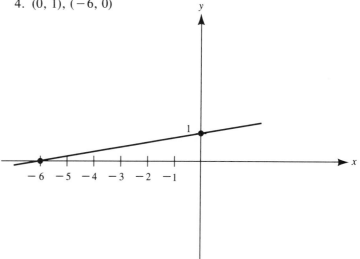

5. (0, 3), (4, 0)

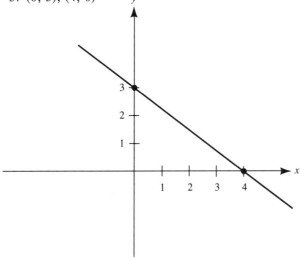

6. (0, 6), (2, 0)

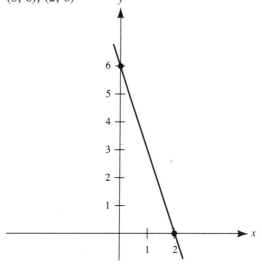

7. $\left(0, \dfrac{9}{2}\right)$, $(-3, 0)$

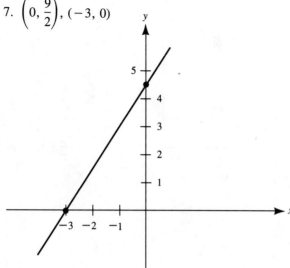

8. $(0, 2)$, $(-4, 0)$

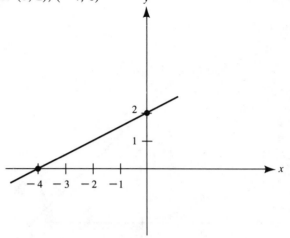

9. $(0, -2)$, $(5, 0)$

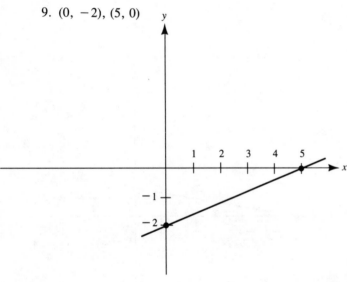

10. $(0, 7)$, $(-2, 0)$

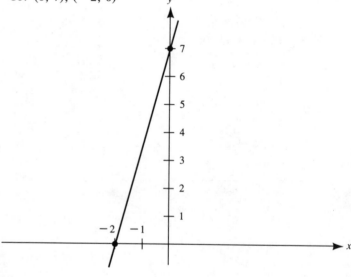

11. $(0, 0)$, $(0, 0)$

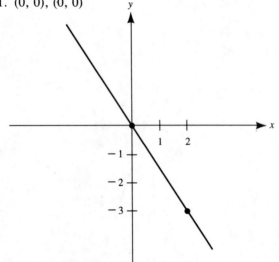

12. $(0, 0)$, $(0, 0)$

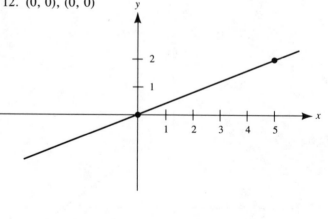

13. (0, 3), no *x*-intercept

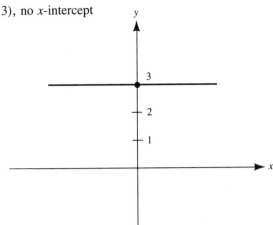

14. $\left(0, -\dfrac{7}{2}\right)$, no *x*-intercept

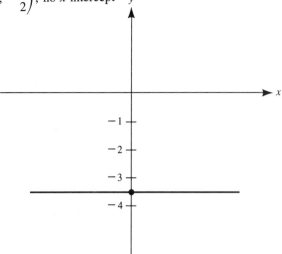

15. No *y*-intercept, $\left(\dfrac{4}{3}, 0\right)$

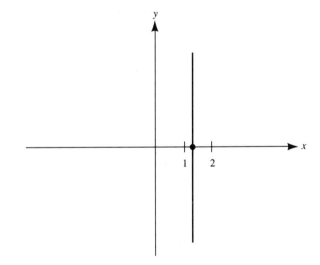

16. No *y*-intercept, $(-2, 0)$

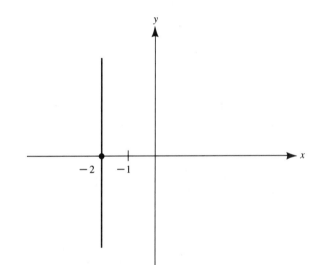

| Exercise 10.4 |

1. 3

2. 2

3. −2

4. $-\dfrac{3}{2}$

5. $-\dfrac{7}{5}$

6. $-\dfrac{5}{4}$

7. −1

8. 2

9. $-\dfrac{1}{2}$

10. $\dfrac{1}{5}$

11. 3

12. $-\dfrac{3}{2}$

13. Undefined

14. 0

15. 0

16. Undefined

Exercise 10.5

1. -4, $(0, -3)$ 2. 5, $(0, -6)$ 3. 2, $\left(0, -\dfrac{4}{3}\right)$ 4. $\dfrac{2}{3}$, $(0, -2)$

5. $-\dfrac{1}{2}$, $\left(0, \dfrac{5}{4}\right)$ 6. -2, $\left(0, -\dfrac{5}{2}\right)$ 7. -2, $\left(0, \dfrac{2}{3}\right)$ 8. $\dfrac{5}{4}$, $(0, -1)$

9. 0, $\left(0, -\dfrac{1}{2}\right)$ 10. 0, $\left(0, \dfrac{5}{6}\right)$

Self-test

1.   (Objective 1)   2.   (Objective 2)

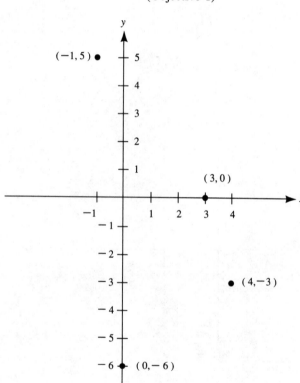

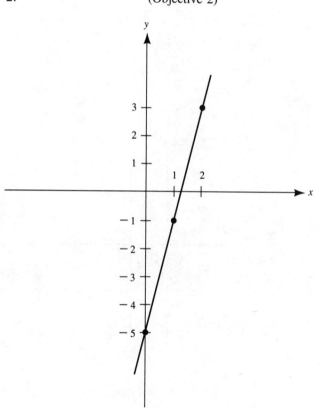

3. (0, 6), (−9, 0)   (Objective 3)

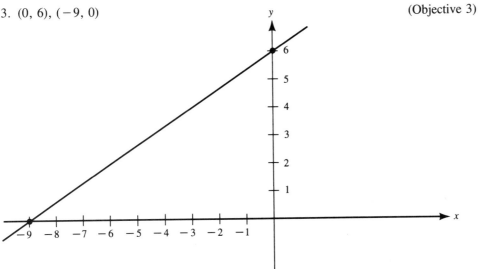

4. $\frac{1}{2}$ (Objective 4)

5. $-\frac{3}{4}, \left(0, -\frac{5}{4}\right)$ (Objective 5)

Unit 11

Self-test

1. $\frac{5}{3}$ (Unit 1)

2. 40 (Unit 4)

3. $11t - 3$ (Unit 6)

4. 4 (Unit 7)

5. −15 (Unit 7)

6. All real numbers (Unit 7)

7. 7.8 inches (Unit 8)

8. 9 cans corn, 15 cans peas (Unit 8)

9. $t = \dfrac{au - s}{a}$ (Unit 9)

10. $(0, -3), \left(\dfrac{9}{2}, 0\right)$ (Unit 10)

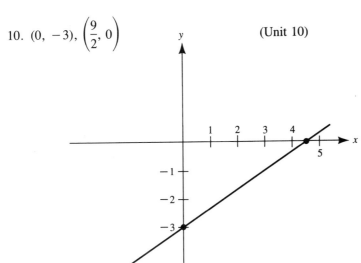

Unit 12

Exercise 12.1

1. (1, 1)

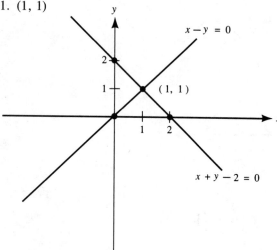

2. (2, −2)

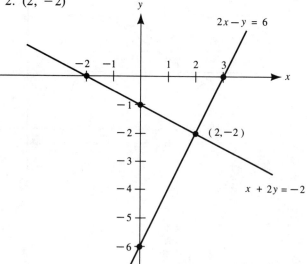

3. (−3, −2)

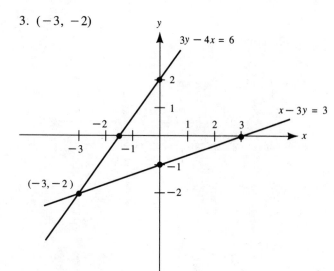

4. (1, −2)

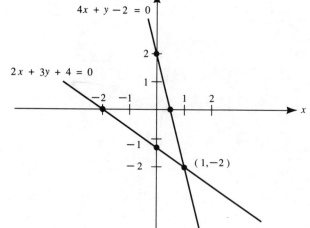

5. (4, 0)

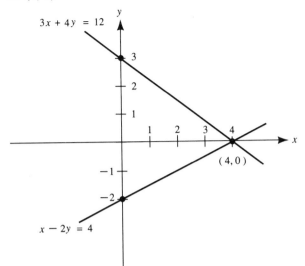

6. (0, −3)

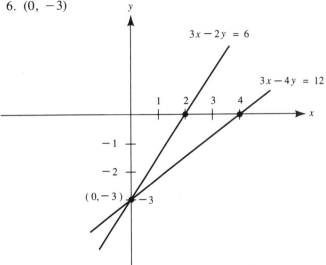

7. (−3, −1)

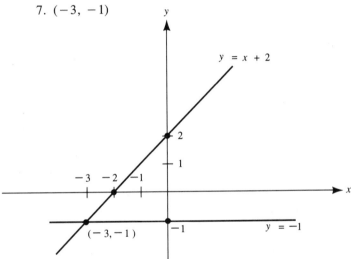

8. (−1, 2)

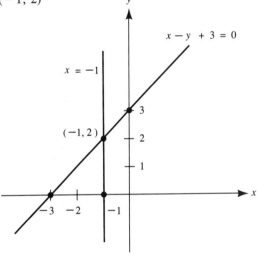

9. Inconsistent

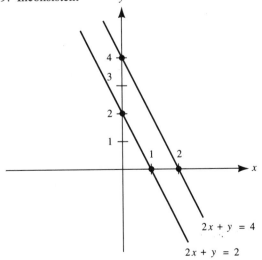

10. Inconsistent

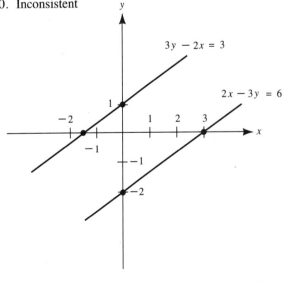

11. Dependent

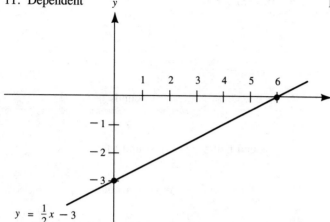

$$y = \frac{1}{2}x - 3$$
$$x - 2y = 6$$

12. Dependent

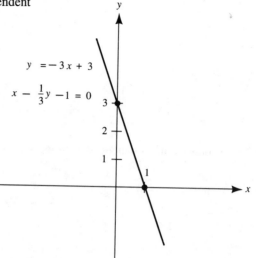

$$y = -3x + 3$$
$$x - \frac{1}{3}y - 1 = 0$$

Exercise 12.2

1. (5, 3)

2. (2, 4)

3. (5, −1)

4. (−3, −1)

5. (2, 3)

6. (4, −3)

7. $\left(2, \frac{1}{3}\right)$

8. (−1, −2)

9. (−2, 3)

10. $\left(\frac{1}{2}, 2\right)$

11. (3, 1)

12. (−3, −4)

13. (2, 5)

14. (0, 0)

15. $\left(\frac{1}{2}, \frac{3}{2}\right)$

16. $\left(-\frac{1}{3}, 3\right)$

17. Inconsistent

18. Inconsistent

19. Dependent

20. Dependent

Exercise 12.3

1. (2, 2)

2. (1, −3)

3. (4, 3)

4. $\left(\frac{1}{2}, -\frac{3}{2}\right)$

5. $\left(-2, \frac{5}{2}\right)$

6. (12, 4)

7. (−1, −9)

8. (−7, −2)

9. (2, −2)

10. (1, 1)

11. (−2, 6)

12. $\left(\frac{1}{3}, 2\right)$

13. (−3, −4)

14. $\left(-\frac{1}{2}, \frac{7}{2}\right)$

15. Dependent

16. Inconsistent

☐ Exercise 12.4

1. 46, 18

2. $16\frac{1}{2}$, $8\frac{1}{2}$

3. $10\frac{1}{2}$ ounces corn, $5\frac{1}{2}$ ounces lima beans

4. 35 pounds of bottles, 15 pounds of cans

5. 24 ounces oil, 8 ounces wine vinegar

6. $\frac{3}{5}$ pound potting soil, $\frac{2}{5}$ pound sand

7. 3 miles per hour, 1 mile per hour

8. 6 miles per hour, 4 miles per hour

9. 440 miles per hour, 40 miles per hour

10. 16 kilometers per hour, 4 kilometers per hour

☐ Self-test

1. $(2, -3)$ (Objective 1)

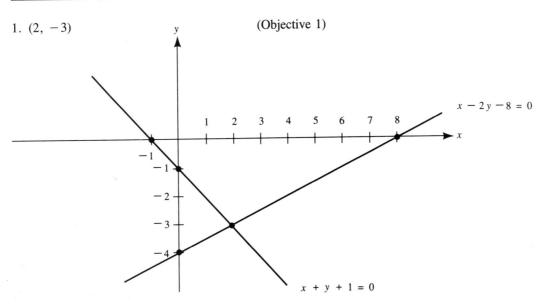

2. $\left(-3, \frac{9}{2}\right)$ (Objective 2)

3. $(-4, -14)$ (Objective 3)

4. $\left(\frac{1}{2}, -\frac{1}{2}\right)$ (Objective 2 or 3)

5. 225 miles per hour, 25 miles per hour   (Objective 4)

☐ Unit 13

☐ Exercise 13.1

1. $x < 4$

2. $x > 4$

3. $x > 10$

4. $x < 2$

5. $x > 3$

6. $x < -15$

7. $x > 1$

8. $x < \frac{5}{2}$

9. $x < 3$

10. $x > -3$

11. $x \leq -\dfrac{1}{3}$

12. $x \geq \dfrac{2}{3}$

13. $x < 1$

14. $x < -1$

15. $x < \dfrac{1}{2}$

16. $x > -2$

17. $x > \dfrac{1}{2}$

18. $x < -4$

19. $x \geq 1$

20. $x \geq 0$

Exercise 13.2

1.

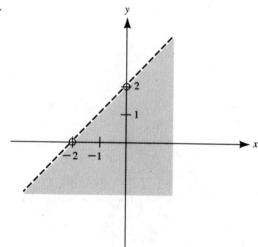

2.

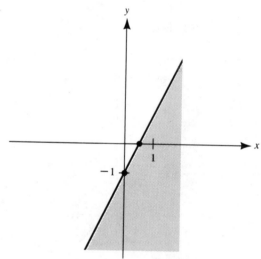

3.

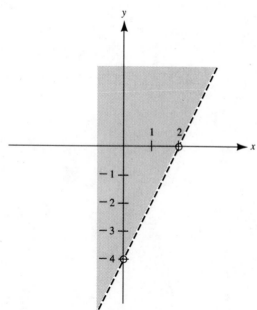

4.

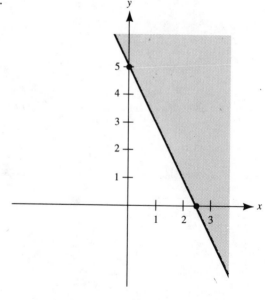

5.

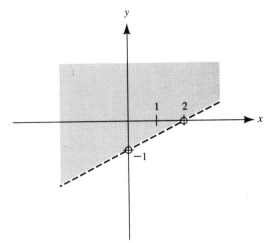

6.

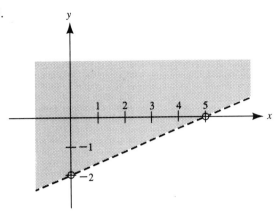

7.

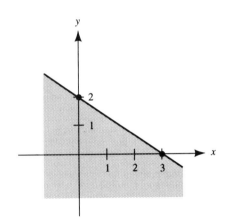

8.

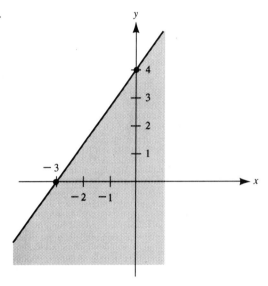

9.

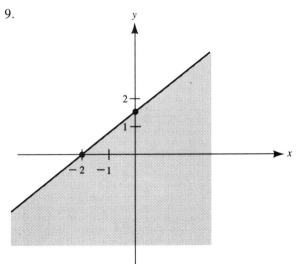

10.

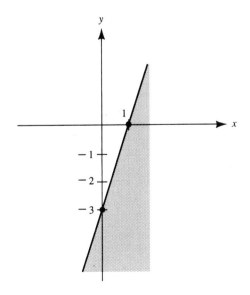

11.

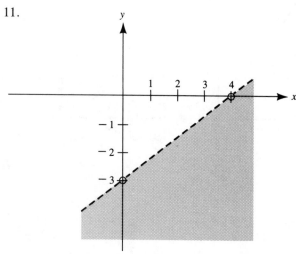

12.

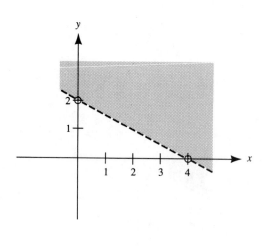

13.

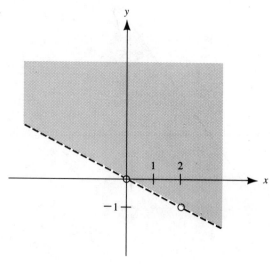

14.

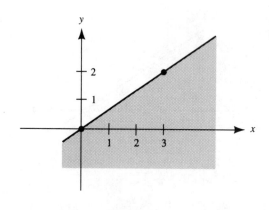

15.

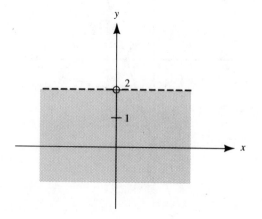

16.

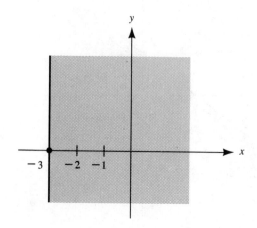

Self-test

1. $x < 5$ (Objective 1)

2. $x \geq 3$ (Objective 1)

3. $x > -\dfrac{7}{3}$ (Objective 1)

4.   (Objective 2)

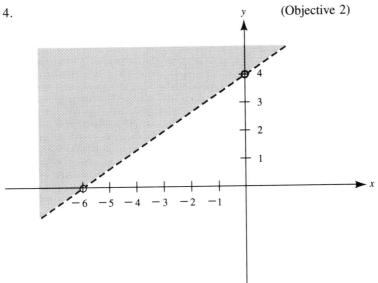

5.   (Objective 2)

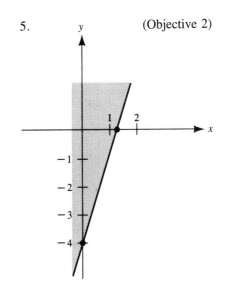

Unit 14

Exercise 14.1

1. 125	2. 243	3. 6	4. -4
5. 16	6. -64	7. -125	8. 144
9. 49	10. -49	11. -27	12. -27
13. -36	14. 8	15. 243	16. -625
17. -1000	18. 1,000,000	19. -1	20. 1
21. -1	22. 1	23. 0	24. 0

Exercise 14.2

1. 25	2. 29	3. -7	4. 60
5. 73	6. 35	7. -189	8. 7
9. 25	10. 16	11. -8	12. -29
13. 28	14. -39	15. 91	16. -89

Exercise 14.3

1. x^8

2. x^{15}

3. Cannot be simplified

4. Cannot be simplified

5. x^4

6. $\dfrac{1}{x^2}$

7. 1

8. x

9. x^{18}

10. x^{10}

11. $x^3 y^3$

12. $\dfrac{x^8}{y^8}$

13. $p^8 q^{12}$

14. $p^{10} q^{12}$

15. $p^3 q^{15}$

16. $p^{15} q^{15}$

17. $2rs^8$

18. $16r^4 s^8$

19. $25r^6 s^2$

20. $5r^6 s^2$

21. $\dfrac{r^2 s^4}{t^6}$

22. $\dfrac{t^6}{r^6}$

23. rs^{11}

24. $\dfrac{1}{rs^3 t^5}$

Exercise 14.4

1. 4500

2. 205,000,000

3. 190,300,000,000

4. 520,000,000,000,000,000,000

5. 0.98

6. 0.00402

7. 0.000001255

8. 0.0000000000000008806

9. 3.2

10. 7.06

11. 2.5×10^4

12. 3.6×10^6

13. 5×10^8

14. 1.06×10^{10}

15. 2.052×10^{12}

16. 2×10^{18}

17. 1.45×10^{-3}

18. 3.02×10^{-4}

19. 5×10^{-6}

20. 9.63×10^{-9}

21. 2.061×10^{-11}

22. 3×10^{-16}

23. 3.72×10^7

24. 9.03×10^{15}

25. 2.24×10^{13}

26. 3.02784×10^{20}

27. 7.82×10^{-7}

28. 7.2×10^{-13}

29. 3.29×10^{-10}

30. 7.64244×10^{-18}

31. 2.70×10^6

32. 1.98×10^{30}

33. 3.94×10^{-22}

34. 1.47×10^{-17}

Self-test

1a. -81

1b. 81   (Objective 1)

2. 92   (Objective 2)

3. $\dfrac{t^5}{r^2}$ (Objective 3)

4a. 5.5×10^9

4b. 6.09×10^{-7} (Objective 4)

5. 2.232×10^{-17} (Objective 4)

▭ Unit 15

▭ Exercise 15.1

1. 6
2. 3
3. 4
4. 3

5. 0
6. 1
7. 4
8. 5

9. 6
10. 4
11. 6
12. 5

13. $x^4 - x^3 - x^2 + x$
14. $-x^2 + 2x + 3$
15. $2x^2 + x - 4$
16. $x^4 - 4x^3 - 3x + 4$

17. $-x^3 + 3x^2y - 3xy^2 + y^3$
18. $x^3 - x^2y^2 + xy$

19. $y^4 - 4xy^3 + 6x^2y^2 - 4x^3y + x^4$
20. $-x^2y^2 + xy + x^3$

▭ Exercise 15.2

1. $2x^2 - 5x - 1$
2. $-x^2 - 4x + 3$
3. $x^2 + x + 5$

4. $3y^3 - y^2 + 4y$
5. $x^2 + 2x + 7$
6. $8y^2 + 10y - 11$

7. $-2z^2 + 2z - 15$
8. $7u^2 - 12$
9. $2x^4 + 2x^3 + 2x^2 - 3$

10. $6y^3 - 4y^2 - 2y - 3$
11. $y^3 - 2y^2 + y$
12. $-10z + 13$

13. $u^2 + 2uv + v^2$
14. $x^3 + x^2y + xy^2 + y^3$
15. $2x^2y^2 - x^2y - xy^2$

16. $s^3 - 5s^2t - st^2 + t^3$

▭ Exercise 15.3

1. $20x^2y^2z^2$
2. $36a^3bc^2$
3. $-5s^3t^3$

4. $-6p^4q^3r^2$
5. $18x^2 - 12$
6. $-10x^2 + 15y^2$

7. $-8x^3 + 2x^2 - 6x$
8. $3x^7 - 12x^5 + 12x^3$
9. $2x^3y^3 - 4x^3y^2 + 2x^2y^2$

10. $x^5y^3 - 3x^4y^4 - 3x^2y^5$
11. $x^2 + 3x + 2$
12. $x^2 + x - 6$

13. $3x^2 - 10x - 8$
14. $4x^2 - 29x + 30$
15. $6x^2 + 5x - 4$

16. $10x^2 - 21x + 8$
17. $x^4 - 8x^2 + 15$
18. $2x^4 - x^2 - 21$

19. $3x^2 + 13xy - 10y^2$
20. $12x^2 - 25xy + 12y^2$
21. $9x^2 + 30x + 25$

22. $4x^2 - 4x + 1$
23. $x^2 - 10x + 25$
24. $16x^2 + 24x + 9$

25. $16x^2 - 9$
26. $36x^2 - 25$
27. $x^3 - 2x^2 - 4x + 8$

28. $4x^3 - 8x^2 + 9x - 9$

29. $x^3 - y^3$

30. $6x^3 - 13x^2y - 3xy^2 + 6y^3$

31. $x^4 - 6x^3 + 13x^2 - 12x + 4$

32. $x^2 - y^2 - 4y - 4$

Exercise 15.4

1. $x + 2y$

2. $2x - 5y$

3. $x - \dfrac{3}{2}$

4. $2x + \dfrac{5}{2}$

5. $xy^2 + 2y + \dfrac{1}{x}$

6. $x - 2y + \dfrac{3y^2}{x}$

7. $x + 2$

8. $x - 2$

9. $x - 3 + \dfrac{-4}{x - 4}$

10. $2x - 1 + \dfrac{3}{x - 2}$

11. $x^2 - 2x + 1$

12. $2x^2 + 3 + \dfrac{1}{2x - 1}$

13. $y^2 + y - 2 + \dfrac{-3}{y - 3}$

14. $2z^2 - 1 + \dfrac{-6}{2z - 3}$

15. $y^2 + 3y + 6 + \dfrac{9}{y - 2}$

16. $x^2 + x + 1$

Self-test

1a. 4

1b. 4   (Objective 1)

2. $8x^2 - 2x - 15$ (Objective 3)

3. $x^3 - 4x^2y + 5xy^2 - 6y^3$ (Objective 3)

4. $6x^2 - xy - 3y^2$ (Objective 2)

5. $2y^2 + 4y + 9 + \dfrac{14}{y - 2}$ (Objective 4)

Unit 16

Self-test

1a. 6

1b. 27   (Unit 4)

2. -3 (Unit 7)

3. $x < -\dfrac{1}{3}$ (Unit 13)

4. $(2, -1)$ (Unit 12)

5. 425 orchestra seats, 120 balcony seats   (Units 8 and 12)

6. $\left(0, \dfrac{3}{2}\right)$, $(3, 0)$, $-\dfrac{1}{2}$ (Unit 10)

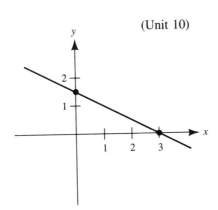

7.   (Unit 13)

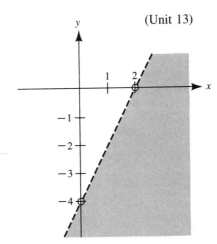

8. 6   (Unit 14)

9. $\dfrac{a^{10}}{b^{10}}$ (Unit 14)

10a. $x^2 - 6xy + 9y^2$

10b. $9x^2 - 16y^2$ (Unit 15)

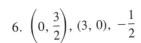

Unit 17

Exercise 17.1

1. $3(x + y)$

2. $2(x - 2)$

3. $3(2x + 1)$

4. $2(x^2 - 3x + 1)$

5. $x(x - 2)$

6. $2x(x + 2)$

7. $2x(y - x)$

8. $4xy(xy + 4)$

9. $x^2y(x - 1)$

10. $xy^2(1 - 3y)$

11. $5(2x^2 - 2x - 1)$

12. $4(3x^2 + 2x - 4)$

13. $x(x^2 + 2x + 2)$

14. $3x(x^2 - 2x - 1)$

15. $xy(x^2 - xy + y^2)$

16. $4x^2y(y^2 + 2y + 3)$

Exercise 17.2

1. $(x + 1)(x - 1)$

2. $(x + 6)(x - 6)$

3. $(2x + 3)(2x - 3)$

4. $(5x + 8)(5x - 8)$

5. $(x + 3y)(x - 3y)$

6. $(7x + 4y)(7x - 4y)$

7. $(4x + 1)(4x - 1)$

8. $(2xy + 1)(2xy - 1)$

9. $(8x + y)(8x - y)$ 10. $(x + 12y)(x - 12y)$ 11. $(x^2 + 2)(x^2 - 2)$ 12. $(7x^2 + 1)(7x^2 - 1)$

13. $(2x^2 + 5y^2)(2x^2 - 5y^2)$

14. $(6x^3 + y^3)(6x^3 - y^3)$

15. $4(2x + 1)(2x - 1)$

16. $y^2(2x + 1)(2x - 1)$

17. $25(2x^2 + y^2)(2x^2 - y^2)$

18. $4x^2(3x + 4)(3x - 4)$

19. $x^2(x^3 + 6)(x^3 - 6)$

20. $16(x^4 + 2y^2)(x^4 - 2y^2)$

Exercise 17.3

1. $(x + 1)^2$ 2. $(x + 4)^2$ 3. $(x - 3)^2$ 4. $(x - 8)^2$

5. $(2x - 5)^2$ 6. $(5x - 4)^2$ 7. $(2x + 1)^2$ 8. $(9x - 1)^2$

9. $(x + 8y)^2$ 10. $(4x - 3y)^2$ 11. $(x^2 - 2)^2$ 12. $(7x^2 + 1)^2$

13. $(x^2 + y^2)^2$ 14. $(3x^3 - 2y^3)^2$ 15. $4(x - 2)^2$ 16. $y^2(2x - 1)^2$

17. $4(x^2 + 3y^2)^2$ 18. $36x^2(x - 3)^2$ 19. $x^4(x^2 + 6)^2$ 20. $16(x^4 - 2y^2)^2$

Exercise 17.4

1. $(x + 1)(x + 4)$ 2. $(x + 3)(x + 6)$ 3. $(x - 3)(x - 4)$ 4. $(x - 2)(x - 8)$

5. $(x + 3)(x - 5)$ 6. $(x + 8)(x - 6)$ 7. $(2x - 3)(x + 2)$ 8. $(3x - 1)(x - 5)$

9. $(x + 5)(4x - 7)$ 10. $(6x - 5)(x + 4)$ 11. $(2x + 3)(3x + 8)$ 12. $(4x + 9)(2x - 5)$

13. $(3x - 2)(3x - 8)$ 14. $(8x + 5)(x - 10)$ 15. $(x + 6)^2$ 16. $(2x - 3)^2$

17. $(4x - 3)(x - 3)$ 18. $(5x + 1)(5x + 4)$ 19. $(x - 6y)^2$ 20. $(x - 3y)(x - 12y)$

21. $3(x + 1)(x + 3)$ 22. $10(x - 2)(x + 3)$ 23. $4(3x - 2)(x - 3)$ 24. $5(2x + 3)(x - 4)$

25. $x^2(x + 1)(x + 2)$ 26. $4x^2(2x - 1)(x + 4)$ 27. $y^2(x + 2)(x - 3)$ 28. $xy(x + y)^2$

29. $(x^2 - 2)(x^2 - 3)$ 30. $(2x^2 - y^2)(x^2 - 2y^2)$

Self-test

1. $(5x - 6)^2$ (Objective 3)

2. $5(x - 2)(x - 3)$ (Objective 4)

3. $3(x^2 - 3x + 9)$ (Objective 1)

4. $(3x + 5y)(3x - 5y)$ (Objective 2)

5. $(2x - 3y)(x + 2y)$ (Objective 4)

Unit 18

Exercise 18.1

1. $0, 5$

2. $0, \dfrac{1}{2}$

3. $0, 3$

4. $0, -3$

5. $0, -\dfrac{3}{2}$

6. $0, \dfrac{1}{3}$

7. $0, -\dfrac{5}{3}$

8. $0, -4$

9. $0, 6$

10. $0, -9$

Exercise 18.2

1. $1, -1$

2. $5, -5$

3. $\dfrac{1}{4}, -\dfrac{1}{4}$

4. $\dfrac{2}{3}, -\dfrac{2}{3}$

5. $2, 3$

6. $-1, -3$

7. $5, -3$

8. $6, -10$

9. $1, -\dfrac{2}{3}$

10. $\dfrac{3}{4}, -1$

11. $\dfrac{3}{2}, -\dfrac{1}{2}$

12. $5, \dfrac{1}{5}$

13. $4, -4$

14. $1, -2$

15. $0, \dfrac{1}{2}$

16. $0, -3$

17. -3

18. 1

19. $4, \dfrac{2}{3}$

20. $-1, \dfrac{1}{2}$

21. $-2, -4$

22. $\dfrac{1}{2}, \dfrac{2}{3}$

23. $-3, 12$

24. $-1, -3$

Exercise 18.3

1. 7 centimeters, 10 centimeters

2. 11 inches, 20 inches

3. 30 feet, 15 feet

4. $7\dfrac{1}{2}$ meters, 5 meters

5. 8 inches

6. 3 centimeters

7. 5 meters

8. 2 feet

Exercise 18.4

1. 5 feet

2. 17 meters

3. 15 centimeters

4. 24 inches

5. 14 inches

6. 9 feet

7. 20 centimeters

8. 60 meters

Self-test

1. $-3, 6$ (Objective 2)

2. $0, 6$ (Objective 1)

3. $4, -\dfrac{1}{2}$ (Objective 2)

4. 26 feet   (Objective 4)

5. 6 centimeters   (Objective 3)

Unit 19

Exercise 19.1

1. $\dfrac{x}{3}$

2. $\dfrac{2x}{3y}$

3. $\dfrac{x}{4}$

4. $\dfrac{5}{x^2}$

5. $\dfrac{x+2}{2}$

6. $1 - 2x$

7. $\dfrac{3+4x}{6}$

8. $\dfrac{3x-4}{2}$

9. $x + 5$

10. $x + 1$

11. $\dfrac{1}{x-10}$

12. $\dfrac{1}{x-2}$

13. $\dfrac{3}{x+3}$

14. $\dfrac{4}{x-3}$

15. $\dfrac{x}{x-4}$

16. $\dfrac{2x}{x+2}$

17. $\dfrac{a}{2}$

18. $\dfrac{x}{2}$

19. $\dfrac{x-2}{x-4}$

20. $\dfrac{x-3}{x+3}$

21. $\dfrac{-1}{x+3}$

22. $3 - x$

23. $\dfrac{2-x}{2+x}$

24. $-(x+1)$

Exercise 19.2

1. $\dfrac{2xy^2}{3z}$

2. $\dfrac{y^2z}{8x^2}$

3. $\dfrac{6x}{y}$

4. $\dfrac{y^2}{2x}$

5. $\dfrac{4}{xy}$

6. $4x^3$

7. $\dfrac{x}{2}$

8. $3x$

9. 1

10. $\dfrac{y+2}{y+1}$

11. $\dfrac{1}{3}$

12. $\dfrac{x}{3}$

13. $\dfrac{2x}{(x-2)^2}$

14. $\dfrac{x}{(x+1)^2}$

15. $-\dfrac{x}{x+1}$

16. $-y$

17. $\dfrac{4x}{(x + 2)(x - 2)}$ 18. $\dfrac{2x + 1}{2(3x + 1)}$ 19. $\dfrac{3x}{(x + 3)(x - 3)}$ 20. $\dfrac{x + 2}{x + 1}$

21. $\dfrac{(x + 3)(x - 2)}{(x + 2)(x - 3)}$ 22. 1 23. -2 24. $\dfrac{3(x - y)}{2(x + y)}$

Exercise 19.3

1. $\dfrac{3x}{x + 3}$ 2. $\dfrac{2x - 4}{x + 4}$ 3. $\dfrac{2x}{x + 1}$ 4. $\dfrac{2}{x + 2}$

5. $x - 1$ 6. $3x$ 7. $\dfrac{2 - 3x}{x^2}$ 8. $\dfrac{x^2 + 4}{x^3}$

9. $\dfrac{2x^2 + 2}{x(x - 2)}$ 10. $\dfrac{x - 8}{x(x + 4)}$ 11. $\dfrac{2x^2 - x + 5}{(x - 1)(x + 1)}$ 12. $\dfrac{2x^2 - 4}{(x + 2)(x - 2)}$

13. $\dfrac{x^2 + 6x - 1}{(x - 2)(x + 1)}$ 14. $\dfrac{x - 8}{(x + 2)(x - 3)}$ 15. $\dfrac{1}{(x + 1)(x + 3)}$ 16. $\dfrac{-1}{(x - 3)(x - 4)}$

17. $\dfrac{1}{(x + 2)(x + 1)}$ 18. $\dfrac{4}{(x + 5)(x + 4)}$ 19. $\dfrac{x + 3}{x - 3}$ 20. 1

21. $\dfrac{x^2}{(x + 4)(x - 4)}$ 22. $\dfrac{-1}{(x + 1)(x - 1)}$

23. $\dfrac{xy}{(x + y)(x - y)}$ 24. $\dfrac{8x}{(x - 4)(x - 4)(x + 4)}$

Self-test

1. $\dfrac{2x}{x + 2}$ (Objective 1)   2. $\dfrac{1}{x(x + 3)}$ (Objective 2)

3. $\dfrac{x - 2}{x + 2}$ (Objective 2)   4. $\dfrac{3x}{(x + 2)(x - 2)}$ (Objective 3)

5. $\dfrac{4x - 8}{(x - 3)(x - 1)}$ (Objective 3)

Unit 20

Exercise 20.1

1. $\dfrac{4}{3}$ 2. $-\dfrac{1}{2}$ 3. -1 4. $\dfrac{3}{2}$

5. $-\dfrac{1}{2}$ 6. 0   7. No solution   8. No solution

9. 4   10. $-\dfrac{3}{2}$ 11. -1 12. -6

13. 7   14. -14 15. 4, 9   16. $-1, 3$

17. $2, \dfrac{2}{3}$ 18. $1, \dfrac{3}{2}$ 19. $\dfrac{2}{3}$ 20. -2

21. 4   22. No solution   23. $1, -1$ 24. -2

Exercise 20.2

1. 20 miles per hour, 10 miles per hour   2. 6 miles per hour, $1\dfrac{1}{2}$ miles per hour

3. 4 miles per hour, 8 miles per hour   4. 6 miles per hour, 16 miles per hour

5. 12 minutes   6. 2 minutes   7. $1\dfrac{7}{8}$ hours   8. $2\dfrac{2}{3}$ hours

9. 12 hours   10. $3\dfrac{1}{3}$ hours   11. 30 minutes   12. 3 hours

Self-test

1. No solution   (Objective 1)   2. $-2, -4$ (Objective 1)

3. -6 (Objective 1)   4. 2   (Objective 1)

5. 9 miles per hour, 3 miles per hour   (Objective 2)

Unit 21

Exercise 21.1

1. 13   2. 15   3. 24   4. 33

5. -8 6. -18 7. Not a real number   8. Not a real number

9. $4\sqrt{2}$ 10. $6\sqrt{2}$ 11. $7\sqrt{2}$ 12. $10\sqrt{5}$

13. $2\sqrt{15}$ 14. $3\sqrt{13}$ 15. $6\sqrt{3}$ 16. $9\sqrt{3}$

17. $\frac{1}{2}$ 18. $\frac{1}{11}$ 19. $\frac{5}{9}$ 20. $\frac{8}{7}$

21. $\frac{1}{6}\sqrt{5}$ 22. $\frac{1}{3}\sqrt{11}$ 23. $\frac{3}{11}\sqrt{2}$ 24. $\frac{6}{5}\sqrt{6}$

Exercise 21.2

1. ± 4 2. ± 7 3. $\pm \sqrt{11}$ 4. $\pm \sqrt{39}$

5. $\pm \dfrac{\sqrt{3}}{4}$ 6. $\pm \dfrac{\sqrt{10}}{3}$ 7. $\pm 5\sqrt{3}$ 8. $\pm 2\sqrt{14}$

9. $\pm \dfrac{3}{2}$ 10. $\pm \dfrac{5}{8}$

11. No real number solution   12. No real number solution

13. ± 5 14. ± 12

15. $\pm \dfrac{2}{5}\sqrt{3}$ 16. $\pm \dfrac{4}{3}\sqrt{2}$

17. $\pm 2\sqrt{6}$ 18. $\pm 2\sqrt{2}$

19. No real number solution   20. No real number solution

Exercise 21.3

1. $2, 3$ 2. $-3, 6$ 3. $\dfrac{1}{2}, 2$ 4. $\dfrac{1}{2}, -\dfrac{3}{2}$

5. $\dfrac{2}{3}, -\dfrac{4}{3}$ 6. $\dfrac{1}{4}, 4$ 7. $2 \pm \sqrt{7}$ 8. $-1 \pm \sqrt{6}$

9. $-2 \pm 2\sqrt{2}$ 10. $3 \pm \sqrt{7}$ 11. $\dfrac{1 \pm \sqrt{3}}{2}$ 12. $\dfrac{-2 \pm \sqrt{31}}{3}$

13. $\dfrac{-3 \pm \sqrt{65}}{4}$ 14. $\dfrac{5 \pm \sqrt{61}}{6}$

15. No real number solution   16. No real number solution

17. $1 \pm \sqrt{3}$ 18. $\dfrac{-1 \pm \sqrt{10}}{3}$

19. $3 \pm 2\sqrt{3}$ 20. No real number solution

1. 2.87 feet, 4.87 feet

2. 0.85 inch, 5.85 inches

3. 1.2 meters, 1.7 meters

4. 1.9 meters, 2.4 meters

5. 1.84 centimeters

6. 2.64 feet

7. 0.7 foot, 2.7 feet

8. 2.9 feet, 5.9 feet

Exercise 21.5

1. (0, −4)
   (2, 0), (−2, 0)
   (0, −4)

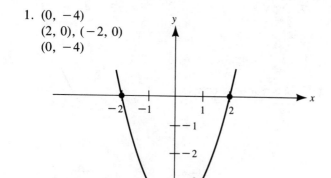

2. (0, −1)
   (1, 0), (−1, 0)
   (0, −1)

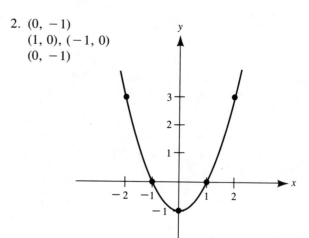

3. (0, 8)
   (2, 0), (4, 0)
   (3, −1)

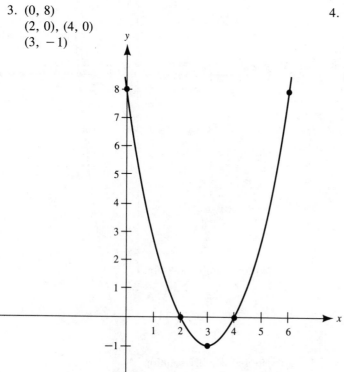

4. (0, −3)
   (−1, 0), (3, 0)
   (1, −4)

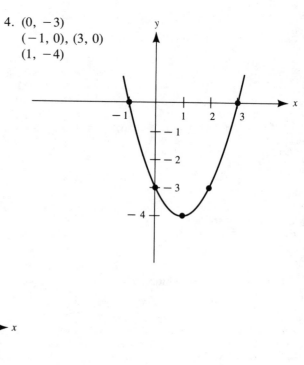

5. (0, −2)
   (−0.7, 0), (2.7, 0)
   (1, −3)

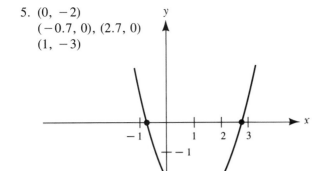

6. (0, −3)
   (4.6, 0), (−0.6, 0)
   (2, −7)

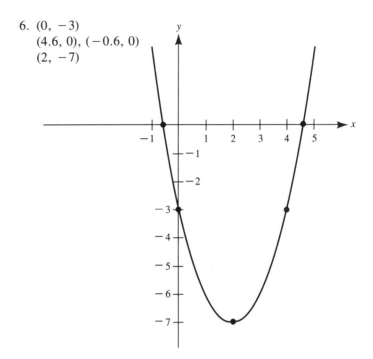

7. (0, −3)
   (1.7, 0), (−1.7, 0)
   (0, −3)

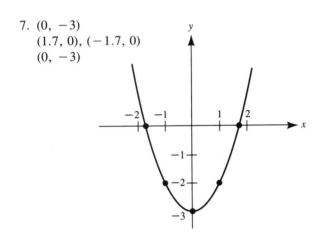

8. (0, −10)
   (2.2, 0), (−2.2, 0)
   (0, −10)

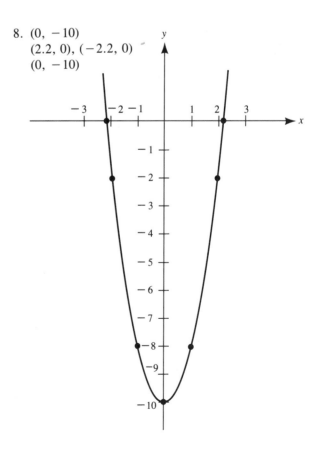

9. (0, 2)
no *x*-intercepts
(1, 1)

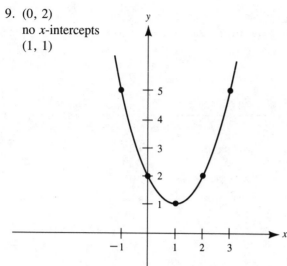

10. (0, 1)
no *x*-intercepts
$\left(-\dfrac{1}{2}, \dfrac{3}{4}\right)$

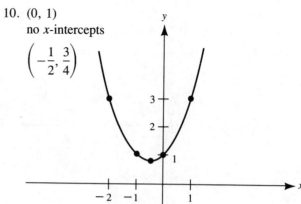

11. (0, 1)
(−1, 0)
(−1, 0)

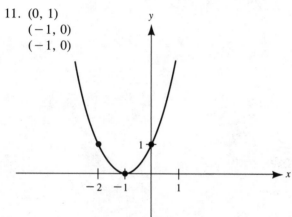

12. (0, 9)
(3, 0)
(3, 0)

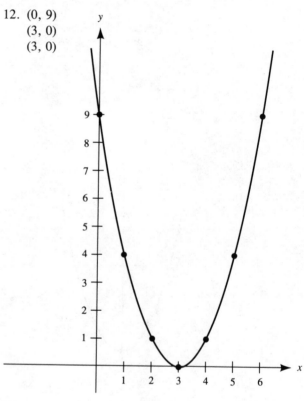

13. (0, 15)
    (−3, 0), (5, 0)
    (1, 16)

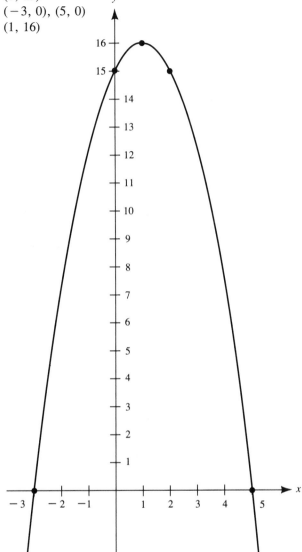

14. (0, 5)
    (−5, 0), (1, 0)
    (−2, 9)

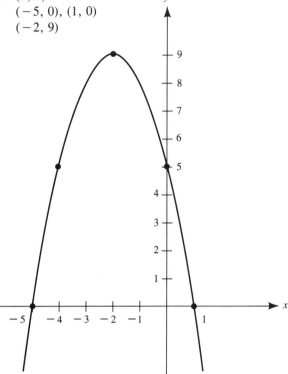

15. (0, 3)
    (−1.8, 0), (0.8, 0)
    $\left(-\dfrac{1}{2}, \dfrac{7}{2}\right)$

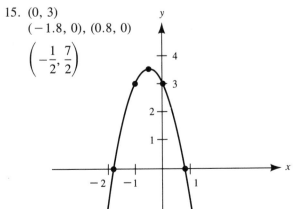

16. (0, −1)
    No x-intercepts
    $\left(\dfrac{1}{2}, -\dfrac{1}{2}\right)$

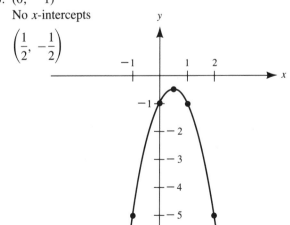

Self-test

1a. $10\sqrt{2}$

1b. $\frac{1}{8}\sqrt{3}$ (Objective 1)

2. $\pm 2\sqrt{2}$ (Objective 2)

3. $\frac{2 \pm 3\sqrt{2}}{2}$ (Objective 3)

4. 2.3 inches, 5.3 inches   (Objective 4)

5. (0, 3)
   $(-1, 0), (-3, 0)$
   $(-2, -1)$

(Objective 5)

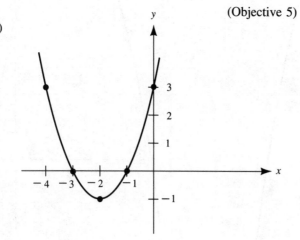

Appendix

Squares and Square Roots

N	N^2	$\sqrt{N}$	$\sqrt{10N}$	N	N^2	$\sqrt{N}$	$\sqrt{10N}$
1	1	1.0 00	3.1 62	51	2 601	7.1 41	22.5 83
2	4	1.4 14	4.4 72	52	2 704	7.2 11	22.8 04
3	9	1.7 32	5.4 77	53	2 809	7.2 80	23.0 22
4	16	2.0 00	6.3 25	54	2 916	7.3 48	23.2 38
5	25	2.2 36	7.0 71	55	3 025	7.4 16	23.4 52
6	36	2.4 49	7.7 46	56	3 136	7.4 83	23.6 64
7	49	2.6 46	8.3 67	57	3 249	7.5 50	23.8 75
8	64	2.8 28	8.9 44	58	3 364	7.6 16	24.0 83
9	81	3.0 00	9.4 87	59	3 481	7.6 81	24.2 90
10	100	3.1 62	10.0 00	60	3 600	7.7 46	24.4 95
11	121	3.3 17	10.4 88	61	3 721	7.8 10	24.6 98
12	144	3.4 64	10.9 54	62	3 844	7.8 74	24.9 00
13	169	3.6 06	11.4 02	63	3 969	7.9 37	25.1 00
14	196	3.7 42	11.8 32	64	4 096	8.0 00	25.2 98
15	225	3.8 73	12.2 47	65	4 225	8.0 62	25.4 95
16	256	4.0 00	12.6 49	66	4 356	8.1 24	25.6 90
17	289	4.1 23	13.0 38	67	4 489	8.1 85	25.8 84
18	324	4.2 43	13.4 16	68	4 624	8.2 46	26.0 77
19	361	4.3 59	13.7 84	69	4 761	8.3 07	26.2 68
20	400	4.4 72	14.1 42	70	4 900	8.3 67	26.4 58
21	441	4.5 83	14.4 91	71	5 041	8.4 26	26.6 46
22	484	4.6 90	14.8 32	72	5 184	8.4 85	26.8 33
23	529	4.7 96	15.1 66	73	5 329	8.5 44	27.0 19
24	576	4.8 99	15.4 92	74	5 476	8.6 02	27.2 03
25	625	5.0 00	15.8 11	75	5 625	8.6 60	27.3 86
26	676	5.0 99	16.1 25	76	5 776	8.7 18	27.5 68
27	729	5.1 96	16.4 32	77	5 929	8.7 75	27.7 49
28	784	5.2 92	16.7 33	78	6 084	8.8 32	27.9 28
29	841	5.3 85	17.0 29	79	6 241	8.8 88	28.1 07

N	N^2	$\sqrt{N}$	$\sqrt{10N}$	N	N^2	$\sqrt{N}$	$\sqrt{10N}$
30	900	5.4 77	17.3 21	80	6 400	8.9 44	28.2 84
31	961	5.5 68	17.6 07	81	6 561	9.0 00	28.4 60
32	1 024	5.6 57	17.8 89	82	6 724	9.0 55	28.6 36
33	1 089	5.7 45	18.1 66	83	6 889	9.1 10	28.8 10
34	1 156	5.8 31	18.4 39	84	7 056	9.1 65	28.9 83
35	1 225	5.9 16	18.7 08	85	7 225	9.2 20	29.1 55
36	1 296	6.0 00	18.9 74	86	7 396	9.2 74	29.3 26
37	1 369	6.0 83	19.2 35	87	7 569	9.3 27	29.4 96
38	1 444	6.1 64	19.4 94	88	7 744	9.3 81	29.6 65
39	1 521	6.2 45	19.7 48	89	7 921	9.4 34	29.8 33
40	1 600	6.3 25	20.0 00	90	8 100	9.4 87	30.0 00
41	1 681	6.4 03	20.2 48	91	8 281	9.5 39	30.1 66
42	1 764	6.4 81	20.4 94	92	8 464	9.5 92	30.3 32
43	1 849	6.5 57	20.7 36	93	8 649	9.6 44	30.4 96
44	1 936	6.6 33	20.9 76	94	8 836	9.6 95	30.6 59
45	2 025	6.7 08	21.2 13	95	9 025	9.7 47	30.8 22
46	2 116	6.7 82	21.4 48	96	9 216	9.7 98	30.9 84
47	2 209	6.8 56	21.6 79	97	9 409	9.8 49	31.1 45
48	2 304	6.9 28	21.9 09	98	9 604	9.8 99	31.3 05
49	2 401	7.0 00	22.1 36	99	9 801	9.9 50	31.4 64
50	2 500	7.0 71	22.3 61	100	10 000	10.0 00	31.6 23
N	N^2	$\sqrt{N}$	$\sqrt{10N}$	N	N^2	$\sqrt{N}$	$\sqrt{10N}$

Index